EUGENE WATKINS MINARD

ONE MOMENT IN ETERNITY — HUMAN EVOLUTION

SECOND EDITION (REVISED)

Order this book online at www.trafford.com
or email orders@trafford.com

Most Trafford titles are also available at major online book retailers.

Note for Librarians: A cataloguing record for this book is available from Library
and Archives Canada at www.collectionscanada.ca/amicus/index-e.html

Library of Congress Control Number: 2005920579

Printed in Victoria, BC, Canada.

ISBN: 978-1-4251-8054-6

Trafford rev. 12/23/2009

www.trafford.com

North America & International
toll-free: 1 888 232 4444 (USA & Canada)
phone: 250 383 6864 • fax: 812 355 4082

ILLUSTRATIONS & CHARTS

TABLE OF CONTENTS

CHRONOLOGY OF CREATION[1]

Years Ago (YA)	Historic Event
10-20 billion YA	Origin of the known Universe
3-4 billion YA	Milky Way Galaxy, our sun and Planet Earth formed
2-3 billion YA	First life forms on Earth
700 million YA	First Great Eocambrian Ice Age
500 million YA	Ice Age in Africa
280 million YA	Ice Age in South America, India Australia, Antarctica
200-140 million YA	Dinosaurs existed for some 60 million years
50 million YA (MYA)	First primates (Great apes)
8-5 MYA	Hominid (ape) ancestors of Man
5-7 MYA	Common ancestor, Chimp and Man
2.5 MYA	Homo habilis (Stone Age Man), Africa
2 MYA	Great Ice Age of Homo habilis
290-140 thousand YA	Homo sapiens (same mitochondrial DNA as humans today)
150-35 thousand YA	Homo sapiens neanderthalensis
100 thousand YA	First anatomically Modern Man
c. 30,000 YA	First Americans via Bering Strait (covered with ice).
10,000 YA	Fourth and most recent Ice Age receding
9,000-4,000 YA	Sumarians in Mesopotamia
5,000 YA	Delta kingdom, Lower Egypt
5,000 YA	Yellow River villages in China

[1] Multiple reference sources

OPENING NOTE

This chronology encapsulates the theme of this book. The process of *human* (H.sapiens sapiens) evolution has occupied only some 100,000 years. This is well within the estimated 10-20 billion years of an evolving universe. 'Tis but a "moment". Can it be human self-extinction or Self-directed (cultural) evolution?

GEORGE SANTAYANA. 1905

Those who cannot remember the past are doomed to repeat it

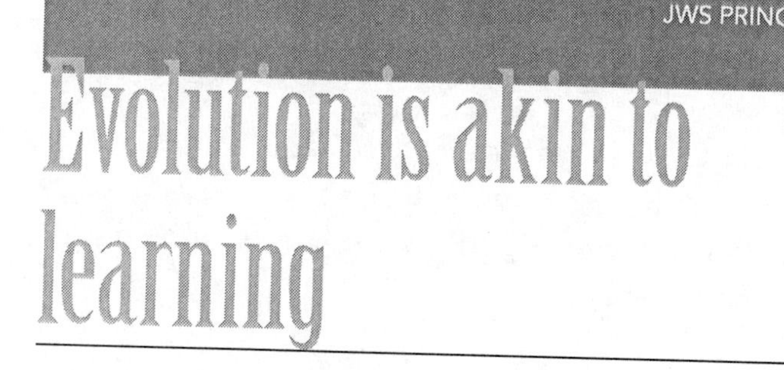

PREFACE

The term, *One Moment in Eternity* could apply to our own galaxy of swirling suns and planets, and to all life on the planet Earth, to the human species and, especially, to each individual's life-time here.

I will quote from my earlier essay, *Evolution of Gods: An Alternative Future for Mankind* (1987):

> In my personal search for understanding of the great mysteries of existence, for a meaning and purpose in life, I have gone from unquestioning faith and inevitable disillusionment to a conviction that only the scientific method can truly enlighten, unite and perpetuate humankind in the face of threatened self-extinction. In a fundamental way, this requires an understanding of *the Great Moving Force* (GMF).

The truth can be very painful, sometimes excruciating. Emotions tend to override logic and the verifiable evidence. I urge my readers to do their own research and then draw their own conclusions.

On approaching old age, many people may wonder what they might have done differently with their "moment in eternity."

I have carefully studied the detailed and sometimes surprising information found in my many referenced sources and believe most to be verifiable when compared with numerous other independent sources. They are part of written history and archaeological discoveries.

A *Chronology of Creation* was provided earlier in this book to help demonstrate the *sweep, magnitude,* and *inter-relationships* of crucial events in the evolution of the universe, planet Earth, life and the human species.

DEDICATION AND ACKNOWLEDGEMENTS

As in my previous essay, *Evolution of Gods* (1987),this book is dedicated to my brother, Oliver Wendell Minard, who died at age 18 in 1944 in England when his bomber crashed and burned. He must not have died in vain.

In January 1943, Wendell, at age 17, and I, at age 18, enlisted in the Army Air Corps. He became the bombardier on a B-24. Due to deafness in one ear, I was transferred to the Army Specialized Training Program (ASTP). When WW II ended in 1945, I was still in the Army as a medical student at Stanford.

My acknowledgements are to Patricia Anderson, Ph.D., my Literary Consultant, to Sarah Kennedy and Elizabeth Bennett, my Author Support Representatives at Trafford Publishing, and to the many authors whose writings provided the vast amount of information used in this present book.

INTRODUCTION

This book is for persons who may be struggling with the same issues which motivated my research and this writing – the great mystery of our personal existence and the universe itself.

The wide variety of subject matter in this book may assist young students in choosing a life-long career, hopefully, in a science and/or history. This could assist them in their own *personal evolution.*

A brief summary of the presently available scientific knowledge regarding evolution is provided here. Evolution is regarded as a *Great Moving Force,* omnipresent, "creating" matter, stars and galaxies, our own star and planet, all life and the human species. An attempt is made to emphasize the almost inconceivable vastness of time and space. This should always remain a source of awe, curiosity, and *humility.* One needs to study astronomy and the biochronology of Planet Earth to begin to appreciate the concept.

Scientific methods are described for dating the stages of evolution of life forms on earth, including humans. Particular attention is given to evolution of the human brain and to our own "personal evolution."

Evolution is as certain as the world is round. Death is as certain as life. We each have one life, one chance, one moment in eternity.

Evolution

CHAPTER ONE

EVOLUTION – THE GREAT MOVING FORCE

THE GREAT MOVING FORCE

[REFS 1-12]

Evolution is a Great Moving Force (GMF), omnipresent, operating at all levels from subatomic to supra-galactic, with simpler forms of matter combining into more complex systems, accompanied by an increase in energy, information, order and adaptability to some larger environments. The process of evolution appears to accelerate – and may be symbolized by the mathematics of the spiral, seen throughout nature, from galaxies to the coriolis effect in hurricanes, to marine mollusks, to the sunflower – and the DNA double helix (1).

In 1951, JWS Pringle compared evolution to *learning*, which accelerates, producing increasing complexity and order. Consider how a small amount of matter and energy, transmitting large amounts of *information* can give control over vast amounts of energy and matter. This is evident in the *blue-prints of life* contained in deoxyribonucleic acid (DNA) and in the formula for conversion of small amounts of matter into the vast energy of a sun, $E = M \times C^2$ (2).

As early as 1936, astronomer Gustaf Stromberg, in his book, *Soul of the Universe*, suggested that the universe might be a "living entity," in which invisible, complex electric *fields* were responsible for guiding the evolution of the cosmos as well as all living things. In 1942, he referred to the research of H.S. Burr at Yale Medical School on living organisms, including the developing embryo (3)(4)(5)(6).

Some scientists have described the universe as an open system, with an external source of energy and matter-energy exchange. All living things are considered to be *open systems.* In contrast, a *closed system,* in which there is no *continuing* energy source, the universe could be expected to evolve, expand or increase in complexity only for as long as there is an energy source or until a *steady state* occurs (7)(8).

From some place and time came a rhyme: "Great fleas have little fleas, upon their backs to bite 'em, and little fleas have lesser fleas, And so on ad infinitum (9)."

Are there smaller basic entities of matter than the known subatomic particles and larger entities than supra-galaxies or even the known universe? The following references discuss subatomic forms of matter at length (10)(11) .

Cosmologists now describe an invisible "dark energy" causing the accelerating expansion of the universe (12).

EVOLUTION OF THE UNIVERSE

[REFS 13-15]

With new technologies, our understanding of the behavior of galaxies, stars and planets has greatly increased. However, the origin and destiny (or fate) of the universe remains a profound mystery. The same is true for the future of mankind.

The beginnings of cosmic expansion has been estimated as ten to twenty *billion* Earth years ago. This event has been referred to as the *Big Bang,* with no acceptable explanation to date. (However, I am reminded of the love duet between Maria and the Captain in the film *Sound of Music*: "Nothing comes from nothing, and *something* never could (come from *nothing*)."

How could something come from nothing? Even *Sub*-sub atomic particles?

It is now certain that galactic clusters are moving apart and accelerating. However, our nearest neighboring galaxy, Andromeda, 2 million light-years away, is *approaching* our Milky Way galaxy. In general, this moving apart requires some mysterious source of energy to counter gravitational attraction and has been compared to raisins in dough in a warm oven.

The further the distance of a galaxy, the greater its speed away from us. Ultimately, objects are moving away at the speed of light. They are then beyond a visible "cosmic event horizon," some 8-12 *billion* light-years away. Galaxies have been photographed up to that distance only. The light we see could have left its source 10 billion light-years ago. A light-year is the distance light travels are 186, 280 miles per second during an Earth year.

Cosmic distances are sometimes described in terms of "Astronomical units" (AUs). An AU is the average distance between Earth and its sun. This is 91.5 million miles or 135 million kilometers.

There is a newer theory that an *eternally existing* universe is inflationary and *self-reproducing*, as if in a chain-reaction, resulting in *fractal-like* universes, growing, *exponentially* in time (13)(14)(15) .

DIMENSIONS OF SPACE AND TIME

[REFS 16-19]

This section is very brief and intended only to inspire someone to seek further understanding. The methods of science continually add to our knowledge of the real world. The vastness of space and time is almost in-conceivable and humbling. [See Figure 1: Cosmic Distances]

Our average-sized galaxy, the Milky Way (Via Galactia), is some 10-20 *billion* Earth-years old, some 100,000 light- years in diameter (a million times that of Earth), and some 1,500 light-years thick. It is a spiraling disc, containing an estimated 100 *billion* stars. Our own star, which we call the sun, classified as a *yellow dwarf*, is far from the center of the galaxy, some 25,000 to 30,000 light-years away, and was formed about 5 *billion* Earth-years ago (16).

Our sun and its planets take about 200 *million* light-years to complete an orbit around the galactic center. The nearest other star in our galaxy is Alpha Centauri, at a distance of 4.2 light-years (17).

If a billiard ball were to represent our sun, the planet Mercury would be at a distance of about 3 meters (from our sun) and Earth at 8.5 meters. The Earth would have a diameter of only 0.5 mm and our moon with 0.1 mm diameter, would circle the Earth in an orbit of 4 cm (18)(19).

[See Figure One: Cosmic Distances]

EVOLUTION OF MATTER

[REF 20]

All matter has evolved from simpler forms, beginning with the first primitive atom of hydrogen, made by fusion of one proton and one neutron. Protons and neutrons were formed from sub-atomic quarks and gluons.

Even today, the smallest atom, hydrogen, makes up over 75 percent of the known universe. All heavier elements of matter have been *fused* in the centers of young stars at incredibly high temperatures and pressures. These temperatures and pressures increase as the core of the young star contracts, causing hydrogen *burning* or fusion, as in a hydrogen bomb, producing the next heaviest and more complex atomic element, helium.

This process continues, step-wise, to produce an end-point and very stable element, iron. No further fusion of larger elements takes place except in the explosions of a nova or super-nova at around 3 *billion* degrees, called a *synchrotron process*. These largest atoms are very unstable and radioactive (20).

The first *molecules* were *diatomic*, from the merger of two atoms. A hydrogen *molecule* consists of two hydrogen atoms. As the primitive universe cooled, larger molecules formed more slowly. The process of molecules *organizing* to become a form of *Life* is presented later under *Life on Earth*.

[See Figure 2: Atomic Evolution]

STARS, GALAXIES AND PLANETS

[REF 21]

Stars evolve. As the intensely hot primordial plasma cools, gravitation causes the condensation and collapse of rotating clouds of hydrogen, which become hydrogen stars, or *first generation stars*, in a young spiral galaxy. This contraction produces, by gravitation, tremendous heat and light. Hydrogen is converted to helium. When the hydrogen is exhausted and helium fusion begins, there is a "helium flash" and a "red giant" star is born.(This is the fate of our own "yellow dwarf" star.) Core contractions and increasing temperatures lead to a "carbon flash", producing oxygen, neon and magnesium. When these nuclear reactions of fusion cease, significant amounts of the star are "shed" to become an immensely dense "white dwarf". Finally, an invisible, cold and "dead" black dwarf. All this takes many billions of years.

Our own medium-sized, rather flat, spiraling galaxy, the Milky Way, contains at least 100 billion stars. Some stars are grouped into associations of hot, super-giant blue-white stars. Born about the same time, they are not bound together by gravity, but are moving apart at high speed. Many other stars are found in *open clusters*, bright and in different colors, moving together through space, bound by gravity. These cluster stars are about the same age. All of the stars in our galaxy's main disc are considered to be younger than Population One" (hydrogen stars).

Our nearest other galaxy is Andromeda (M31), about the same size and spiral shape as our Milky Way. It is over 2 million light-years away and moving closer. It will not collide with our galaxy for another 2 billion years.

Galaxies cluster together. Our Milky Way is part of the "Local Group" of some 35 galaxies, across some 3 million light-years of space (21).

[See Figure 3: Cosmic Spirals]

Far, far beyond the Andomeda galaxy is the Hydra cluster, at 2 or 3 billion light-years away. At the so-called "cosmic horizon" where no more light can be seen, stars are moving away at the speed of light. The universe is expanding and accelerating !

EARTH IN TIME

[REFS 22,23]

To further grasp the time scale for Earth and life in the Universe, imagine compressing the whole time from the beginning of the universe, about 15 *billion* years into a single 24-hour day.

A few seconds after midnight, stable atoms begin forming from sub-atomic particles. Hours later, in the early dawn, stars and galaxies appear. Only around 6 P.M., our solar system was forming. By 8 P.M., the first forms of life are seen and by 10:30 P.M., the first vertebrates crawl onto the land. Dinosaurs roam from 10:30 P.M. until 4 minutes before midnight. Ten seconds before midnight, our first human-like ancestors walk upright. The Industrial Revolution (1750-1850 CE) and subsequent events occupy less than one-thousandth of a a second before midnight. (Note here that since the Industrial Revolution began around 1750

CE/AD, the Earth's environment has changed more than in the previous 4.5 *billion* years.)

About 225 million years ago, one giant land mass, now referred to as Gondwana, began to break up into what is now South America, Africa and India, Antarctica and Australia. All were separated from North America and Eurasia (Laurasia). As the Earth's continents drifted, the earth's magnetic poles shifted. Some animal species were isolated for very long periods until new land bridges formed. Incidentally, Earth's North magnetic pole in Canada is moving Northwest, rather rapidly, towards Siberia.

There is ample proof that Earth's magnetic poles have reversed many times during the last half-billion years, coinciding with the extinction of many species (22)(23).

LIFE ON EARTH

[REFS 24-26]

Life on Earth could only occur when our medium-sized yellow dwarf star, following its birth 5 billion years ago, had stabilized its internal hydrogen-to-helium fusion. Some 4.6 billion years ago (BYA), Earth became a planet, condensing by self-gravitation from our sun's spinning halo of gas and dust.

Tremendous internal temperatures and pressures caused the escape of lighter elements such as hydrogen and helium, while a molten core of heavier metals such as iron and nickel formed. Around 2 BYA, the Earth's crust had cooled to the boiling point of water, which then could condense into a primitive atmosphere containing ammonia, methane and carbon dioxide. Water boiled at the equator, and it could rain at the poles.

Only after another billion years a "secondary" atmosphere rich in oxygen appeared, but without a protective ozone layer.

Volcanic activity was intense. Oceans and continents were

forming. In the past billion years the continents drifted and the magnetic poles shifted. Magnolias grew in Iceland, with coral in Arctic seas, while glaciers covered what is now Brazil and the Congo (24)(25).

Modern plants and animals requiring oxygen were now able to emerge some 600 to 700 million years ago (MYA). By that time there was a protective ozone layer in Earth's atmosphere. In Earth's primitive atmosphere of ammonia, methane, water and carbon dioxide, intense ultraviolet (UV) radiation could have produced the first simple organic molecules, as accomplished in labs with electricity through gas mixtures. "Life" can be defined as involving self-replication, as well as self-regulation, which requires complex molecules. On planet Earth, the first complex molecules were evolving an estimated 4 billion years ago (BYA), primitive cells 3.6 BYA and the first *nucleated* cells only about 1.5 billion years ago (26).

[See Figure 4:Geologic Time]

EVOLUTION OF LIFE

[REFS 27-28]

By some mysterious process, smaller molecules *organized themselves* into larger and more complex molecules, "macro-molecules." Some of these molecules, such as phospholipids, could organize themselves into a membrane, which can in turn enclose a cell.

The most primitive cells, found in 3.5 billion year old rocks, had no nucleus and deoxyribonucleic acid (DNA) had not yet "clustered" into discrete chromosomes. These most primitive cells are called *prokaryotes,* in contrast to all cells in multi-cellular forms of life today, called *eukaryotes*. These latter cells do contain chromosomes, nuclei and other organelles such as *mitochondria*. Only prokaryotes existed through the Pre-Cambrian Archean Eon (3.9-2.5 billions of years ago). These lasted some 2.5 billion years predating the Cambrian Period (before 600 million years ago, where present day eukaryocyte cell types existed (27).

The modern-day eukaryocyte is believed to have been formed by the inclusion of a bacterium or tiny primitive cell-like structure call a *mitochondrion.*

This new addition allowed cells to derive energy from complex compounds ("food") by oxidation. For the evolution of mitochondria, see the previous reference.

Some 600 million years ago (MYA), at the beginning of the Cambrian Period, free-living single celled organisms began to spend part of their life-cycle in colonies, as is seen today in hydra and volax. It must have been a small step to a permanent colony, a multi-cellular plant or animal. This suggests some kind of continuing mystery regarding "self-organizing systems," directing evolution at all levels supra-galactic to sub-atomic (28).

EXTRA-TERRESTRIAL ORIGINS OF LIFE

[REFS 29-31]

Meteorites have been found to contain precursors of biological compounds such as carbonaceous chondrites. These compounds have amino acids with equal left-handed and right-handed structures. Amino acids in living organisms on Earth have left-handed configuration. In addition, meteorites contain a higher ratio of carbon-13 to carbon-12 than is found in any living creature on Earth. One more clue has been the detection in outer space of compounds representing intermediate steps in the formation of amino acids. Extra-terrestrial sources of these compounds do not alone explain the process by which living systems have become self-replicating by way of DNA and RNA (29).

According to Rhawn Joseph, there are rather compelling arguments for an extra-terrestrial source of the "seeds of life" by way of meteorites, even from the beginning of life on Earth. The generally accepted process by which atoms become molecules, which become cells, which become multi-cellular animals or plants has been questioned. Joseph believes that, as an environment acts on gene selection, a "collection of like-minded cells" begins to interact, forming a nerve net, then congregating in an anterior head region, forming a primitive brain (30).

Joseph's theory of *evolutionary metamorphosis* attempts to explain why different species of "humanity" have possibly overlapped in time and appeared simultaneously in different parts of the world. His theory states that DNA had to come from space and other planets in other galaxies and had to be widely distributed on Earth, awaiting a proper environment. He seems to endorse the "visitations" of extra-terrestrial beings although he considers most to be hoaxes. He concludes that we have the genetic potential to be "gods." The question of Homo sapiens arising in one place only or in many places is discussed in the following reference and the preponderance of evidence supports a single place of origin in Africa (31).

DATING THE EVOLUTION OF NEW SPECIES

[REFS 32-36]

The gradual and sequential change in plants and animals during the past 3 billion years has been dated by three methods (32):

1) Radiometric dating by the known radioactive decay rates of certain elements found in sedimentary rocks;
2) The sequence of rock strata with distinctive fossils; (plant and animal remains) known to be abundant at a particular geologic time, e.g., fish were first abundant in Devonian rocks from 350-400 million years ago;
3) Changes in the Paleomagnetic pole positions from the remnant magnetism of iron-oxide particles in rocks and sediments.

The chronological sequence, hereditary or genealogical lineage, of various species of animals has been most clearly demonstrated by the study of *molecular genetics* or biochemical similarities (33)(34)(35).

These studies have mostly involved nucleotide sequences in genes rather than the amino acid sequences of their products. They have shown the close relationship between monkeys, primates and great apes as well as the time of their divergence into different species and the changes in their genes during the past 50 million years. {Incidentally, a *species* can be defined as a group with specific genes, capable of interbreeding).

Comparisons between nucleic acids as well as proteins (e.g., hemoglobin), various tissues and anti-serum have given similar results. Immunity tests with anti-human serum show 100 percent reactions by all humans and, revealing lineage distances, have shown 64 percent by gorillas, 42 percent by orangutans, 29 percent by baboons, 10 percent by sheep, 2 percent by horses and no reaction by kangaroos, indicating the latter as the most distant common ancestor among the species tested.

Comparison of chromosomes of African apes (chimpanzees and gorillas) show very little difference from humans, who show twice the difference from Asian apes. This is what had been suspected on the basis of anatomical and other comparisons. Based upon these studies, humans had a common ancestor with the chimpanzee 5-7 million years ago (MYA), the gorilla about 10 MYA, orangutan 16 MYA, Old World monkeys 35 MYA and New World monkeys 45 MYA.

Since the mapping of the human genome and comparing it with other animal species, there has been some controversy regarding the possibility of interbreeding species. Of course, part of the definition of a *species* is that only members can interbreed.

A recent novel concerning this subject has obviously involved extensive research by the author, who states in the beginning – "This novel is fiction, except for the parts that aren't (36) ."

[See Figure 5: Young Chimpanzee]

HUMAN EVOLUTION

[REFS 37,38]

Fossils of the oldest human-like ancestor are probably those of the African Proconsul, dating to about 20 MYA. Fossils of the most ancient "human" genus, Homo, date to about 2 MYA. Based mainly upon brain volume (1,100 to 1,200 cubic centimeters), the first Homo sapiens fossils date to about 200,000 years ago. Our most recent direct ancestors were *Homo sapiens neanderthalensis,* with fossils found in Europe and Iraq, dating to between 50,000 to 100,000 years ago. Their brain volumes averaged 1,300 to 1,500 cubic centimeters, the same as modern humans.

Our own present species of *Homo sapiens sapiens* appeared in various parts of the world about 40,000 years ago (YA). In this very short cosmic time, no significant *physical* evolution has occurred. This includes *no change* in *instinctive* behaviors. Any further human evolution will have to be *cultural* (37)(38).

Human behavior is driven by ancient *instincts,* as with all forms of life. The *personal-survival instinct* causes each human to avoid harm or death. It is the basis for the persistent and fervent belief in some kind of *life-after-death* regardless of the lack of any kind of *Scientific,* verifiable, evidence for this. The *species-survival instinct,* also very powerful, causes all living things to reproduce. Even the common house fly avoids danger and "mates" – and then multiply in an almost logarithmic manner.

[See Figure 5: The Chimpanzee]

BRAIN EVOLUTION

[REFS 39,40]

One of the most profound of all mysteries is how the complete human body, and the brain in particular, can develop form a single fertilized microscopic egg cell.

The modern human brain, contained in the small space of some 1,500 cc, weighing about 3 pounds, can reflect the universe. It is surely the most complex and magnificent "creation" of the Great Moving Force (evolution), other than the universe itself. The human brain is more flexible than previously thought and has powers of regeneration not fully appreciated until recently. The number of brain *neurons* are estimated at 100,000,000,000 or 100 billion and the number of supporting *glial cells* at more than 10 trillion (10,000,000,000,000). Compare these numbers with the estimated 100 billion stars in our Milky Way galaxy (39)(40).

Can you imagine the Internet or World Wide Web (WWW) having as many interconnections ? [See Figure 6: Primate Brains]

Equal to the supreme mystery of the Universe itself, is *HOW* – not *WHY* – the merging of two microscopic cells (sperm and egg) can produce the magnificent living body of each mammal on planet Earth ! [See Figure 3 – Cosmic Spirals: Galaxies, Hurricanes *and* DNA]

PERSONAL EVOLUTION

(ZERO REFS]

Personal evolution is accomplished by the acquisition and creative use of knowledge. Cultural evolution of the human species occurs by the personal evolution of its individual members.

The *individual* human, during their very short *cosmic* lifetime, may evolve through the same principle as the universe and their own species. This is by learning and adapting to the laws of *nature,* and to become a creative and caring member of their species and a guardian of their only home – planet Earth. Hopefully, they may have had more joy than sorrow by the end of their "One Moment in Eternity".

As J.W.S. Pringle said in 1951, "Evolution is akin to learning". As with learning, evolution tends to accelerate.

Through learning the *secrets of nature*, most reliably through the methods of Science, the individual may become the most healthy and creative and caring individual possible. They must always feel a need to doubt and, again, ask "How" – not "Why".

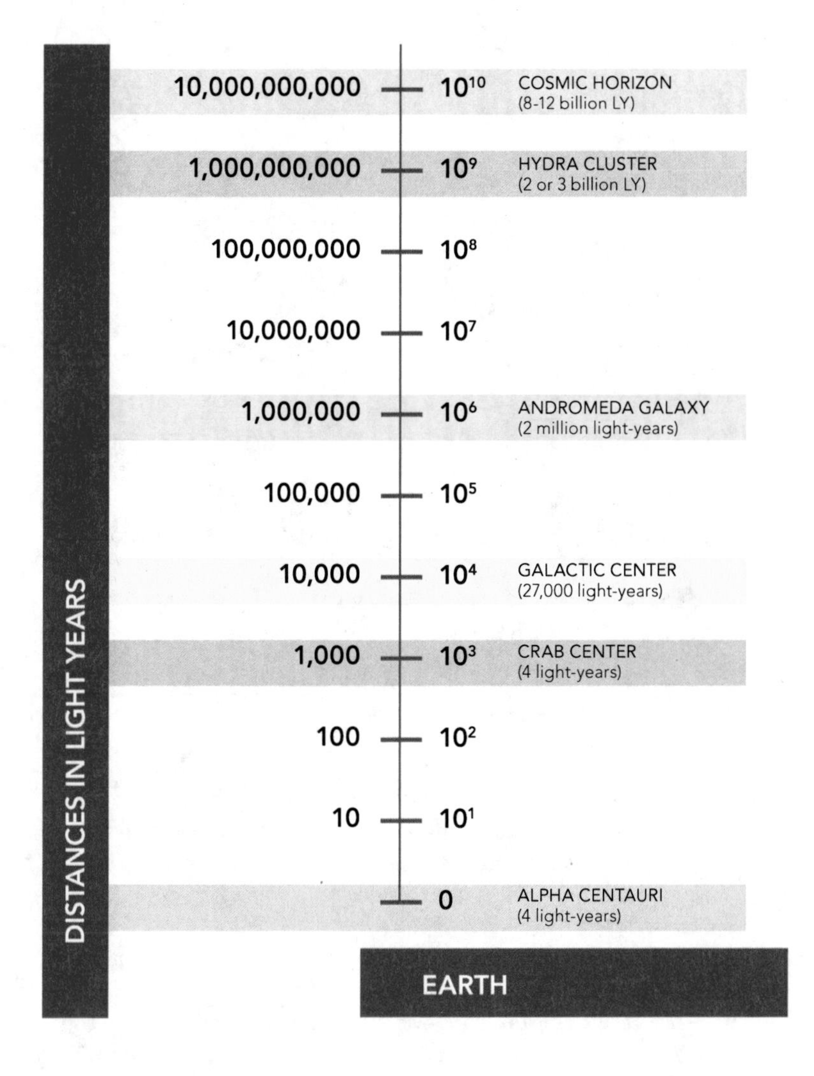

FIGURE 1: COSMIC DISTANCES
(From Earth to the cosmic horizon)

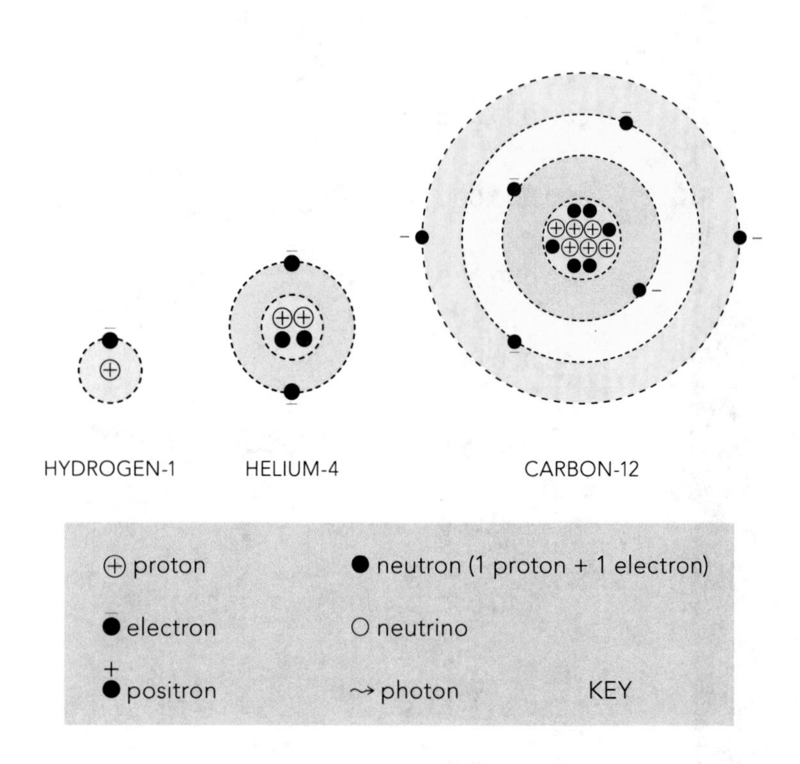

All matter has evolved from simpler forms beginning with hydrogen, which makes up more than 75% of the known universe.

This most simple atom is created by the fusion of one proton and one neutron – which in turn consist of sub-atomic quark and gluon particles.

All heavier elements are produced by fusion in the center of young stars at extreme temperatures and pressures.

This very complex step-wise process, beginning with the conversion of hydrogen to helium, proceeds to an end-point with iron, the heaviest element.

The only further fusion process occurs in the explosion of nova or super-nova at some 3 billion degrees. This is called a synchrotron process.

FIGURE 2: ATOMIC EVOLUTION
DATA FROM MULTIPLE SOURCES INCLUDING
www.astrophysicsspectator.com/topics/stars/FusionHydrogen.html

WHIRLPOOL GALAXY
(Image from the Wikopedia Free Encyclopedia]

HURRICANE
(Courtesy NASA)

FIGURE 3: COSMIC SPIRALS
(Galaxies, Hurricanes – and DNA)

Time Scale (eon)	Era	Period	Epoch	Millions of Years Before Present (approx.)	Duration in Millions of Years (approx.)	Some Major Organic Events
Phanerozoic	Cenozoic	Quaternary	Recent (last 5,000 years)		1.6	Appearance of humans
			Pleistocene	1.64		
		Tertiary	Pliocene	5.2	3.5	Dominance of mammals and birds
			Miocene	23.5	18.3	Proliferation of bony fishes (teleosts)
			Oligocene	34	10.5	Rise of modern groups of mammals and invertebrates
			Eocene	55	21	Dominance of flowering plants
			Paleocene	65	10	Radiation of primitive mammals
	Mesozoic	Cretaceous		146	81	First flowering plants Extinction of dinosaurs
		Jurassic		208	62	Rise of giant dinosaurs Appearance of first birds
		Triassic		245	37	Development of conifer plants
	Paleozoic	Permian		290	45	Proliferation of reptiles Extinction of many early forms (invertebrates)
		Carboniferous	Pennsylvania	320	30	Appearance of early reptiles
			Mississippian	363	43	Development of amphibians and insects
		Devonian		409	46	Rise of fishes First land vertebrates
		Silurian		459	30	First land plants and land invertebrates
		Ordovician		505	66	Dominance of invertebrates First vertebrates
		Cambrian		545	40	Sharp increase in fossils of invertebrate phyla
Precambrian	Proterozoic	Upper		900	355	Appearance of multicellular organisms
		Middle		1,600	700	Appearance of eukaryotic cells
		Lower		2,500	900	Appearance of planktonic prokaryotes
	Archean			3,900	1,400	Appearance of sedimentary rocks, stromatolites, and benthic prokaryotes
	Hadean			4,500	600	From the formation of Earth until first appearance of sedimentary rocks; no observable fossil organisms

**Note:* Dates derived mostly from Harland et al. Some geologists divide the Precambrian eon into two major eras, Proterozoic and Archean, and then denote the Hadean as the first Archean period (Fig. 9-13). However, the exact dates that mark each geological period are often only approximate, and other authors provide somewhat different time spans.

FIGURE 4: GEOLOGIC TIME
(Adapted from *EVOLUTION – 2000* by permission of Jones & Bartlett Publishers, Boston)

FIGURE 5: YOUNG CHIMPANZEE
OUR COMMON ANCESTOR WAS SOME 5-7 MYA
(By permission of National Geographic –
Michael Nichols, photographer)

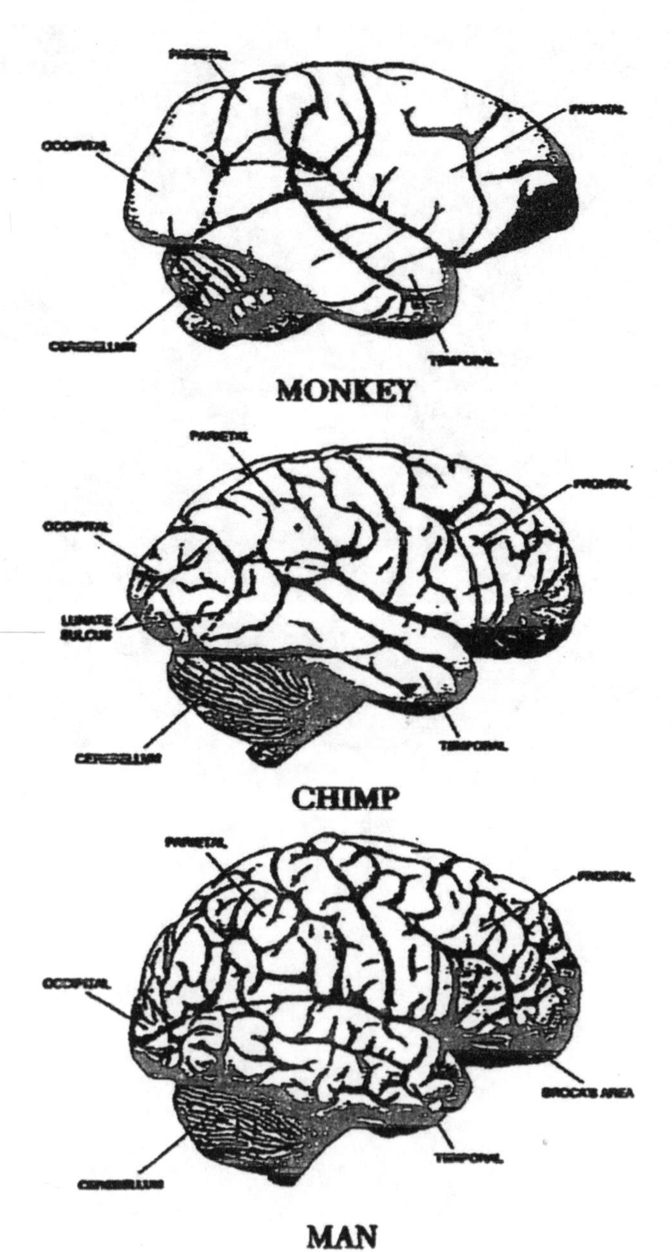

FIGURE 6: PRIMATE BRAINS
(Adapted from July 1974 Scientific American with permission from the artist, Alan Eiselin. Authors were Holloway and Prentiss – "The Casts of Fossil Hominoid Brains")

REFERENCES AND NOTATIONS: CHAPTER ONE

THE GREAT MOVING FORCE

1) Davis, Philip. *Spirals – from Theodorus to Chaos.* Wellesley, MA. AK Peters, Ltd., 1993. [Considerable History and math. Especially see page 202. ALSO see: Livio,Mario.*The Golden Ratio.* New York. Broadway Books, 2002. *Spirals,* pp.116-120. ALSO see: *Geometry is all* – in The Economist, November 24,2007]
2) Pringle, JWS. *On the parallel between learning and evolution.* In Buckley, W. (Editor). *Systems Research for the Behavioral Scientist.* Chicago. Aldine, 1968.
3) Stromberg ,Gustaf. *The Soul of the Universe.* North Hollywood, California. Educational Research Institute edition,1965.Original publication by David McKay Company,1940.
4) Burr,Harold Saxton. *The Fields of Life: Our Links with the The Universe.* New York. Ballantine Books, 1972. 5)Becker,Robert;Seldon,Gary. *The Body Electric: Electro-magnetism and the Foundation of Life.* New York. William Morrow and Company, 1985.
6) Berra,Tim. *Evolution and the Myth of Creationism.* Stanford, CA. Stanford University Press, 1990.
7) Chiasson,Eric. *Cosmic Evolution: The Rise of Complexity in Nature.* Cambridge, MA, and London. Harvard University Press, 2001. See pages 26-29.
8) Gribbin,John. *Genesis – The Origins of Man and the Universe.* New York. Dell Publishing, 1981.
9) Swift,Jonathan. *Poems II.* 651, 1733.
10) Charap,John. *Explaining the Universe: The New Age of Physics.* Princeton, NJ. Princeton University Press, 2002.[See page 85 for "fleas on fleas" and Chapter 8: *Microcosm.*]
11) Maddox,John. *What Remains to be Discovered.* New York, NY. Touchstone Books/Simon and Schuster, 1998.
12) *Scientific American Book of the Cosmos.* David Levy, (Editor). New York. St. Martin's Press, 2000.[See Liss, Tony; Trypton,Paul. *The Discovery of the Top Quark.* Rubin, Vera. *Dark Matter in the Universe.* Peebles, Turner and Kron. *The Evolution of the Universe.*]

EVOLUTION OF THE UNIVERSE

13) Linde, Andrei. *The Self-Reproducing Universe.* Found in *Scientific American Book of the Cosmos.* Levy, David (Editor). New York. St. Martin's Press, 2000.

14) Schroeder, Manfred. *Fractals, Chaos and Power Laws: Minutes from an Infinite Paradise.* New York: W.H. Freeman and Company, 1991.[See page 152: *The Clustering of Poverty and Galaxies.*]

15) Mendelson, Jonathan; Blementhal, Elana. *Chaos Theory* and Fractals. Found at www.mathjmendl.org/chaos

DIMENSIONS OF SPACE AND TIME

16) Gutsch,William. *1001 Things Everyone Should Know About The Universe.* Doubleday,1998. [See Chapter 11: *Stellar Geriatrics.*]

17) Gribbin, John. *Genesis: The Origins of Man and the Universe.* New York. Dell Publishing, 1981.

18) Shklovskii,I.S.; Sagan,Carl. *Intelligent Life in the Universe.*New York. Dell, 1966. [See pages 28-29]

19) Hawking,Stephen W. *The Theory of Everything: The Origin and Fate of the Universe.* Beverly Hills, CA. New Millenium Press,2002. [See *the Boundary Conditions of the Universe,* pages 136-140]

EVOLUTION OF MATTER

20) Davies,Paul. *The Cosmic Blueprint: New Discoveries in Nature's Creative Ability to Order the Universe.* New York. Simon and Schuster, 1998.Chapter 9:*The Unfolding Universe.*[ALSO see: Mason,Stephen. *Chemical Evolution.* Oxford. Clarendon Press,1991.]

STARS, GALAXIES AND PLANETS

21) Davis,Philip. *Spirals – From Theodorus to Chaos.* Ellesley,MA. AK Peters, 1993. [ALSO see: *The Cosmic Life Cycle – Origins of the Universe.* Scientific American. 2007.]

EARTH IN TIME

22) Strickenberger, Monroe. *Evolution – Third Edition.* Boston, MA. Jones and Bartlett,2000.[*Biological effects of drift,* pages 102-107]

23) Becker,Robert;Selden,Gary. *The Body Electric: Electro-magnetism and the Foundation of Life.* New York. William Morrow,1985.

LIFE ON EARTH

24) en.wikipedia.org/wiki/Geologic_timescale. 25)Eicher,Don. *Geologic Time-Second Edition* .Englewood Cliffs, NJ.Prentice Hall,1968. Pages 52,61.

26) Miller,James Grier. *Living Systems.* New York. McGraw Hill, 1978. [See definition of "living systems", page 1027]

EVOLUTION OF LIFE

27) Strickberger,Monroe. *Evolution-Third Edition,* 2000. [See Figure 9-13 on page 174 and Table 6-2 on page 93 and pages 180-183 regarding mitochondria]

28) Ward,Mark. *Beyond Chaos: The Underlying Theory Behind Life,the Universe and Everything.* New York. St. Martin's Press,2001.[See pages 86,112,175 and 278 regarding "self - organizing systems"]

EXTRA-TERRESTRIAL ORIGINS OF LIFE ON EARTH

29) Stanley,Steven. *Earth and Life in Time – Second Edition.* New York. W.H. Freeman,1989.[See page 259.] 30)Joseph, Rhawn. *Astrobiology.* San Jose, CA. University Press, 2001.

31) Strickenberger,Monroe.*Evolution – Third Edition,*2000. [Page 486.]

DATING THE EVOLUTION OF NEW SPECIES

32) Pilbeam,D. *The descent of hominoids and hominids.* Scientific American, March 1984.

33) Strickenberger,Monroe. *Evolution – Third Edition,*2000. [Chapter 20: *Primate Evolution and Human origins.* Also see Pages 465,468-480]

34) Strickenberger.Monroe. *Evolution – Third Edition,*2000. [Chapter 25: *Cultural Evolution.* ALSO see pages 48,486,488 regarding migrations out of Africa from 200,000 years ago.]

35) Roberts,J.M. *Ancient History.* London, UK. Duncan Baird, 2004. [Chapter 2: *Homo* sapiens]

36) Crichton,Michael. *Next.* New York. HarperCollins, 2006. [NOTATION: Although this is a novel – perhaps science fiction – the author's very extensive research of the legitimate scientific literature on the subject of inter-breeding between species justifies further consideration]

HUMAN EVOLUTION

37) Gilbert,Scott. *Developmental Biology.* Sunderland, MA. Sinauer,1997.

38) Buchanan,Mark. *Nexus.* New York W.W. Norton,2002.

BRAIN EVOLUTION

39) Strickberger,Monroe. *Evolution – Third Edition,*2000. [This 721 page textbook is full of very informative illustrations, and statistics. The ultimate source regarding *Evolution*]

40) Maddox,John. *What Remains to be Discovered: Mapping The Secrets of the Universe, the Origins of Life, and the Future of the Human Race.* New York. Touchstone- Simon and Schuster, 1998.[NOTATION: This

434 page textbook can be used for self study –like *Science 101* – and is full of surprises]

PERSONAL EVOLUTION

PART TWO

HUMAN NATURE

CHAPTER TWO

HUMAN NATURE

HUMAN NATURE OVERVIEW

[REFS 1-87]

Humans (*homo sapiens sapiens)* have continued in a transition stage of evolution since their appearance as a species somewhere beyond 100,000 years ago. They are without doubt the most cruel and destructive, but also the most compassionate and creative of all animals on this planet.

They have almost unlimited creative potential – *if* they continue to evolve, rather than joining the vast majority of once living species (1).

Homo sapiens sapiens approached extinction several times during their 100,000 years – mostly from natural causes beyond their control. However, the world is now a different place in the 21st century. Humans now probably do have the power to join the other extinct species, through their own ignorance and greed.

Human nature, gone astray, can be described in terms of monumental greed, causing *crimes against humanity,* and profound ignorance, causing a persistent *blind faith* in man-made religions and gods, self-proclaimed prophets or saviors. Historically, insatiable lust for power has brought wars of conquest, often in the guise of *religious zeal.*

Greed is innate and based on the instinct for survival, physical or emotional. Monumental *greed* motivates wars of conquest, corrupt governments and businesses.

There are many more *fools* (meaning ignorant, deceived or *fooled persons*, not *retarded*) than *criminals*, but criminals often use *fools* to achieve their goals. Human *instincts* don't change. They are in the DNA of our species. However, *ignorance* can be overcome by learning the provable *laws of nature*. Unfortunately, many world citizens do not yet have access to the *scientific* information which contradicts most of the *religious* interpretations of history. This *religious world-view continues* to prevail among the majority of the planet's six and one half billion humans.

For many humans, belief in the *supernatural* continues without realizing that the amazing human conquest of space and time and control of the many devastating disease pandemics has been made possible by the discovery of *natural laws* through the methods of science, not by religion. Today, *religious wars* continue to threaten human survival. Basically, all wars are about *greed* and *power.* They have always involved *crimes-against-humanity.*

BASIC INSTINCTS

Most humans, like all other animals, struggle to survive and to reproduce. These behaviors are motivated by these two most *Basic Instincts: Self-survival* and Species-*survival.*

The Self-survival instinct clearly accounts for the historical human denial of the *permanency* of death.

The Species-survival instinct is manifested by the powerful need for sexual relations between males and females – as with all animals.

A number of very destructive human behaviors do not seem explained by either of these basic instincts alone. They will be discussed later under *Self-extinction Behaviors.*

PERSONAL-SURVIVAL INSTINCTS

The varieties and possible *perversions* of the *personal* survival instinct need much more study. This very powerful source of human behavior has historically seemed to have been the basis for:

1) The ancient and persistent belief in a *life-after-death.*
2) Struggle for food and shelter and safety.
3) Curiosity, Creativity and Compassion.
4) Self-defense against *Monsters and Madmen.* (Some of these have been difficult to identify easily and early).

LIFE-AFTER-DEATH *MYTH*

- Titus Lucretius Carus (c.60 BCE): Mind *and* soul are mortal. When one is dissolved, so is the other.
- Rene Descartes (1596-1650): *Cogito, ergo sum* (I think, therefore I am).
- Eugene Minard (2008): For an eternity I did not exist. Too soon, I will again cease to exist – forever.

Although humans have long been aware of the inevitability of their personal death, their need to survive has made them vulnerable to deliberate deception and to charlatans. The latter are persons who have claimed to communicate with the angels or gods. They have claimed to have received various *Rules* of behavior which will result in some kind of *heaven or hell in* some kind of *life-after-death.*

Based upon the personal-survival instinct, this belief continues widely today, despite the *total absence* of any scientific evidence for its reality – quite the opposite. It continues partly due to childhood indoctrination and in the unavailability of scientific or historic knowledge for many of the world's citizens. It is really about *greed* and *power* over the persons deceived.These rules from some imaginary *God* have been the excuse for *Holy Wars* and many horrific *Crimes-against-humanity* – the same as today.

Actually, human *men* have made all the rules or laws concerning human behavior. Too often these conflict with the *Laws-of-Nature,* which are un-changeable. Natural *laws* continue to be discovered by the methods of Science. No *Holy scripture* ever revealed the planet Earth's total insignificance in a seemingly endless Universe – nor how to put men on the moon.

Possibly the best arguments regarding the persistence of this *life-after-death* wish was written and published in 2006 by acclaimed author Richard Dawkins (2).

CURIOSITY, CREATIVITY AND COMPASSION

Curiosity, creativity *and* compassion are basic aspects of the Self-survival instinct.

Even before the major discoveries and scientific inventions, humans built beautiful edifices and created works of art – music in particular (Consider Johann Strauss, Victor Herbert, Rudolf Friml, Franz Lehar and many others). Many early humans had also shown compassion for their fellow humans. Compassion and *empathy* are surely based upon *both* the Personal *and* Species survival instincts.

SELF-DEFENSE AGAINST *MONSTERS* AND *MADMEN*

Another example of both the Personal *and* Species survival instinct surely includes the moments in history when the innocent victims of tyrants rose against them. Some such actions have been recorded as *Revolutions.*

HISTORIC REVOLUTIONS

Revolutions as rebellions against tyranny, in self defense, are discussed by causes, outcomes and dates in most books on world history. Someof the best known and with far-reaching consequences include: the Protestant Revolution, the American and French and Russian Revolutions (3)(4)(5).

SPECIES-SURVIVAL (SEXUAL) INSTINCTS

Human sexual behaviors are determined partly by the species DNA and partly by the social and religious culture that individuals are born into. Often these seriously conflict with the instinctive *Laws-of-Nature.*

However, these Species-survival (sexual) instincts may be equal to, or sometimes more powerful than, the rules of behavior produced by the restrictions of a certain culture or religion. In cases of rape or murder to conceal a sex crime, the punishment may knowingly involve life in prison or *death-row.*

SEXUAL BEHAVIORS IN HISTORY

Important clues to these *Laws of Nature* can be found in the study the sexual behaviors of humans and other animals from pre-historic times (6-18).

A better understanding is needed regarding the difference between sexual behaviors as *nature* obviously has "intended" and the behaviors condemned and punished according to rules invented by men.

Some insight regarding sexual behaviors of human males may be found in a historical novel about the *Geisha* of Japan during the great global depression of the 1930s (18).

Other sexual behaviors which need research and may be on the increase include: wife/husband *swapping* (for sex), incest, unwanted pregnancies, sexually transmitted diseases, prostitution, homosexuality and sex with non-human animals.

EMANCIPATION OF WOMEN

For thousands of years, many – if not most – females have had the status of virtual slaves to men. This has been manifested by multiple wives, concubines, mistresses, harems and prostitutes.

However, within the past century, almost startling changes in regard to female *equality* with males have occurred in the *demo-*

cratic (*non-theocratic*)countries. The most obvious exception has been in countries still dominated by the Islamic religion.

In the United States in 1920, the 19th Amendment to the U.S. Constitution allowed women to vote !! During the 1920's, women could properly show their ankles, and more – in public.

After World War Two, for awhile, mini-skirts were the rage. Nude females could soon be viewed in popular magazines such as *Playboy.*

By 1993 the Internet (World Wide Web) became available. Then, all restrictions on opinions and information seemed to evaporate – except in countries where the Internet was not legal. In recent years, this made all kinds of *pornography* easily available. Almost all that was needed was to type the word *Sex* and hit the *Search* button.

One major change, with questionable outcomes, has been the Internet access of photos and videos showing nude women and girls(allegedly at least age 18) in photos or videos in various sexual activities. These young *models* are from many countries around the world. The motivation seems to be for money – and for *pleasure,* perhaps.

An article in the February 2005 Readers Digest titled *No Strings Sex,* described how TV shows, such as *Sex and the City, Spring Break and Girls Gone Wild,* had influenced girls 12-17 years old. They were now twice as likely to start having sexual intercourse as those who watched very little – per a RAND Corporation and University of California study. Also noted was that 15-19 year old females have gonorrhea more than any other age group. "Internet hookups with strangers are part of every parent's nightmare."

A major problem involves female health and safety. Research is clearly needed to determine what behaviors are harmful or dangerous, even to *informed and competent* females who participate. Clearly, the graphic portrayal of physical pain or injury – real or pretended – could be a basis for criminal prosecution. Enforceable laws are needed on an international level. According to the current national media, very little progress – or interest – is evident.

Research is especially needed regarding the possible serious *emotional* harm to females of any age when there is any use of force, threat or coercion by a potential sex partner. Physical force, e.g. rape is a crime-against-humanity. This certainly should include mutilation of female genitalia (FGM)- as a cultural phenomenon over hundreds of years, especially in Africa. It continues on a wide scale today – resisting prolonged efforts by the United Nations and non-governmental agencies.

Another most abominable crime-against-humanity is world-wide child prostitution. These last two topics are continued below under Self-extinction Behaviors/*Crimes–against-humanity.*

ADOLESCENT SEXUAL BEHAVIORS

A lengthy article is available regarding apparent recent changes in the sexual behaviors of adolescents in the United States, Britain, Canada, and India (19).

Briefly, female children in the U.S. are entering puberty at least two years earlier than previous generations. This may mean that they are ready physically for sex, but not emotionally or cognitively – surely not ready for motherhood.

According to the American Academy of Pediatrics this has created a "major public health problem."

The U.S. president, George W. Bush, had insisted on a funding focus of "abstinence only" and using foreign aid to pressure the *end* of condom education in third-world countries.Teens in the U.S. have reported that the *media* ranks second only to school sex education programs in providing information. Much from TV and films seems "misleading and unrealistic", but often accepted as factual by teen-agers.

Girls who "hang out" with older boys are more likely to be pressured into sex, to get sexually transmitted disease and unwanted pregnancies. Oral sex has been reported as becoming more popular than vaginal intercourse. Drugs and/or alcohol are increasingly used before sexual relations.

In Britain, a survey in 2006 found that more adolescents were waiting before having sexual intercourse. The largest drop was among 14 and 15 year olds. However, of all the *Western* countries, Britain reportedly has the highest incidence of teenage pregnancy and increasing incidence of sexually transmitted disease. These include the biggest rise in syphilis. However, Chlamydia, genital warts and herpes were most common.

In Canada, research has shown that girls with poor self-esteem were more likely to engage in sex by the end of seventh grade – around age 11.

In India, sex outside marriage is "not uncommon", but girls were influenced by what their girl friends were doing.

Misconceptions were common regarding prevention of pregnancy or diseases.

On March 11,2008, the U.S Center for Disease Control(CDC) reported that one in four female adolescents were infected with at lease one sexually transmitted infection.[www.cdc.gov]

LEGAL AGE FOR CONSENT

The legal age of consent for sexual relations varies worldwide. In most countries this seems to be between ages 14 and 16. Countries where the minimum legal age for male-female sex is 18 include Egypt, El Salvador, Guatemala, Haiti, Iraq, Peru, Swaziland, Turkey, Viet Nam. The youngest age in any country is listed as 13 in Argentina, Canada (12,14,18), Cyprus (13,17),Japan (13,18), Mexico(12,18), Nigeria (13), South Korea, Spain, Syria (13,15).

In the United States, the legal age is 18 in Arizona, California,

North Dakota, Oregon, Tennessee, Virginia, and Wisconsin. In all other states the age varies between 14 and 16. In some states, parental consent is required – if under 18 (20).

SELF-EXTINCTION BEHAVIORS

Self-extinction behaviors concern some human activities which do not exclusively relate to either *self*-survival or *species*-survival.

Human self-extinction behaviors must originate partly from *both* the species DNA (inborn instincts) and that which has been learned (cultural).

These *Self-extinction* behaviors includethe continuing cruel and *destructive* wars, always accompanied by *crimes-against-humanity*. Crimes-against-humanity include genocide and massacres, *human-trafficking*, slavery, torture, rape and child prostitution.

ENDLESS WARS AND ASSOCIATED ATROCITIES

[REFS 21-26]

Wars and their associated atrocities are human behaviors evidenced by archaeological and written records. There seems no conceivable way that *WAR*, except in self-defense, can contribute to the survival of the human species.

There is a long history in most cultures of *atrocities*, in peacetime or war. They have often involved torture before death (21).

Wars have usually been motivated by personal *greed* and *lust*. Too often they have hidden behind the guise of religion, even though they included the invading and looting of other countries, with rape and the taking of slaves.

To quote Steven LeBlanc, an archaeologist at Harvard University: "To understand much of today's war, we must see it as a common and almost universal human behavior that has been with us as we went from ape to human." "The chimp and human behaviors are almost completely parallel." The author's book is very wide-ranging, historically and geographically (22)."

After World War Two it became very clear that civilian deaths far exceeded those of military personnel. Many died in concentration or prison camps and from the massive bombing of cities. These included the cruel and needless destruction by the bombing of Dresden, Frankfurt, Hamburg, Tokyo, Hiroshima and Nagasaki. The same massive toll on civilians occurred in Korea and Viet Nam. These actions, mostly by the United States, clearly were in violation of the Geneva Conventions of 1977 and the added Protocol of August 12, 1949 (23)(24)(25)(26).

The United States invasions of Afghanistan (2001) and Iraq (April 9, 2003) have again resulted in many more civilian deaths than military.

The historical record demands a conclusion that all wars of aggression have been based upon unabashed *greed*, even when under the guise of religious proselytizing. Young men kill other young men they have never met and have no personal reason to hate. The only winners are the arms makers and money lenders.

CRIMES AGAINST HUMANITY

Crimes-against-humanity have been defined at Special Criminal Tribunals and by the United Nations General Assembly.

These now must include *genocide, torture, rape, human trafficking and child prostitution*. Not yet so defined by UN Criminal tribunals – but greatly needed – is female genital mutilation (FGM)(27)(28). Very little progress has been made in prosecuting and punishing these global problems.

GENOCIDE AS A CRIME-AGAINST-HUMANITY

Genocide, in the 20th century, was defined as a *Crime-against-humanity* by the General Assembly of the United Nations on

December 11,1946. Article I stated that "...genocide, whether committed in time of peace or in time of war, is a crime under international law."

Article II continued: "... genocide means any ... acts committed with intent to destroy, in whole or part, any national, ethnical, racial or religious group . . ."

Before the United Nations General Assembly established the International Criminal Court (ICC), effective July 1,2002, there was an incident-by-incident approach to international crimes by Special Tribunals. This began with the Nuremberg War Crimes trials against Nazi defendants in 1945 and against some 28 alleged Japanese war criminals between 1946-1947.

The Nazi Holocaust between 1933-1945 resulted in some six million civilian deaths (29).

At Nuremberg, criminal charges against Germans included *aggressive warfare, extermination of civilian populations (Jews especially),widespread use of slave labor, looting of occupied countries, maltreatment and murder of prisoners of war.* In Tokyo, the charges were similar (30).

A more recent example of on-going *genocide* is in the Darfur region of the Sudan. Beginning in 2003, several African tribes (known as the Sudan Liberation Movement) protested the Arab-dominated Sudan government. The government backed *militia* called *janjaweed* (horsemen) began burning villages, attacking residents and raping females.

Despite repeated condemnations by the United Nations and the then U.S. Secretary of State, Colin Powell, calling this *genocide,* the bombing, burning, raping, killing continued. In 2006, the UN Security Council approved replacing 7,000 African Union forces with 20,000 UN peacekeepers. The Sudan government refused the non-African forces as *neo-colonists.*

Only in early July 2008 have charges of genocide been made

against the Sudanese government by the International Criminal Court. An arrest warrant was issued for Sudanese president Omar Hassan al-Bashir.

In 2007, the UN accused the government of *crimes- against-humanity,* with an estimated 2 million persons displaced and some 450,000 killed. Some blame was also placed on *rebels,* also accused of rape and torture of civilians.

In December 2007, the U.S. Congress sent a Bill to Presi-dent Gorge W. Bush. This Bill would *promote or* authorize States, local governments and *"other investors"* to remove Sudan-related assets from their portfolios.

The genocide continues. For a detailed account of the historic suffering of these people, see TIME Almanac- 2006: *Chad,* pages 745-746. (Note that Chad oil fields may be a hidden agenda.) Also see *The worsening chaos of Darfur.* The Economist, October 13, 2007.

GENOCIDES IN HISTORY

Genocide surely should include the atrocities of the Catholic Inquisitions between 1231-1834 AD/CE (In this book, see Chapter Four).

How much of this group behavior resides in the human species DNA and how much is learned or "cultural ?"

Many massacres, like genocide, have occurred other times in history. In the past century, the most deadly examples of genocide and/or massacres – have included the following few:

- Cambodia under Pol Pot, between 1975-1979, after the end of the Viet Nam War, some 2 million persons died from executions and starvation (25-30 % of the population) (31)(32)(33).
- Serbia-Herzegovina genocide, between 1992-1995,resulted in some 20,000 deaths (33).
- In Rwanda, on April 6, 1994, the apparent assassination of

> President Habyarimana, a Hutu, resulted in an immediate door-to-door systematic killing of thousands of *Tutsis* by the armed forces and *Hutu* militia. The UN peacekeeping force only *observed*, because their *mandate* was only to *monitor.*

The second day, ten Belgian UN soldiers, while guarding the *Hutu* prime minister, were tortured and killed – presumably by Tutsis. That same day, the Tutsis *Patriotic Front* launched a major offensive to end the killing and to rescue 600 of its troops surrounded by Hutus.

On April 9th, French and Belgian troops rescued their citizens and Americans – but no Rwandan employees of Western governments. On that date, the International Red Cross estimated "tens of thousands" of Rwandans had been murdered.

On April 14th, the UN Security Council voted to withdraw most of their troops. A secret report by the U.S State Department called the killings *genocide* (34).

Civil wars in Africa have continued for years. Some of the countries involved include Sierra Leone, the Democratic Republic of Congo (child soldiers committing atrocities), Liberia (Charles Taylor and diamond conflicts with Angola),Uganda (The Lord's Resistance Army),and, especially Darfur in Sudan (35)(36)(37).

The prevention of genocide seems far in the future, since the authority of the International Criminal Court is not yet accepted by all nations – even though they are members of the United Nations General Assembly. This includes the United States.

The number of massacres in history could be a surprise and a shock – especially for those who know little about world history. These are described in a *List-of-massacres,* available on the Internet. These date from 334 BC/BCE to World War II. The *List* provides overwhelming written and archeological evidence of the depravity of many members of the human species – over many millennia (38).

Crimes-against-humanity now must also include *torture, rape, human trafficking and child prostitution.*

TORTURE AND RAPE

Torture of the innocent – with or without murder – is a human behavior almost beyond understanding. It is difficult to explain except as a very perverted or very misguided *instinct.*

This sadistic group behavior has often occurred in connection with *genocide, mass murder* and other *crimes against humanity.* The latter term was defined by an *International Criminal Tribunal* and now includes *rape* (39) .

Torture, historically, has seemed most evident when used in relation to religious dissension, whether as punishment or to obtain a confession. In some form it has been used by the United States government with suspected *terrorists,* especially since September 11, 2001.

Historical descriptions of torture, in horrifying detail, may be found in three references (40)(41)(42).

Rape, like genocide, has recently been declared a *Crime-against-humanity.* It has been described as ".. a spoil of combat." Also, " Rape has been considered a war crime for years." "It falls under a definition of torture." "Crimes against humanity – per the International Court (ICC) Statutes – include rape, sexual slavery, enforced prostitution, enforced sterilization, whether during war or peace, if of a widespread or systematic nature (43)."

The International Criminal Tribunal for the Former Yugoslavia on February 22, 2001, declared Rape to be a *crime against humanity* and also a form of *enslavement.*

Three Bosnian Serbs were convicted for their involvement in "rape camps" in Foca. These began operating in 1992. Some females were as young as 12 years. Some victims reported being assaulted by up to 10-15 soldiers for hours at a time (44).

There are many historical records of rape in war time.

One event in the past century must include the few weeks in 1937-1938, when the Japanese Army invaded Nanking, China, and began an orgy of torture and massive killing unsurpassed in history – in such a short time period (45).(Note that the Japanese military were also provided with "comfort women", mostly Korean, during WW II.)

Other incidents of mass murder and rape by German military and, later, Russian military, are recorded in several books (46)(47)(48)(49).

More recent sexual atrocities against women and girls have been occurring in Darfur, Sudan, for over 4 years now. The perpetrators are the Arab militia called *janjaweed* . They have been assisted by the Muslim Sudanese government. The government denies any such incidents, in disregard for the African Union, European Union and the League of Arab States denouncing this as a conspiracy between the Arab militants and the Sudanese government (50).

As of March 2008, massive numbers of women and girls continued to be raped and severely injured by combatants in the Congo Civil War – for over 10 years now. In the opinion of the author of this book, the Code of Hammurabi (2500 BC/BCE) should be enforced by international law. The perpetrators must be brought to justice in the International Criminal Court, using large UN peacekeeping forces for their apprehension.

The Hammurabi Code – referred to as the Law of Talon – means the same penalty for the offender as the crime they committed.

HUMAN TRAFFICKING

This international crime has been discussed in detail by the newspaper PRAVDA. This involves the global problem of forcing children into prostitution and the dangers to young Russian women recruited for jobs in other countries, but ending up as prostitutes and virtual slaves.

In Vladivostok about 40 *clandestine* brothels reportedly *supply* underage prostitutes of *both sexes* as young as 14 years (51).

CHILDREN AS PROSTITUTES

"Southeast Asia is thought to be the hotbed of child prostitution." According to UNICEF (the United Nations Children's Fund), more than one million children in Asia including Thailand, Philippines, and India, have been sold to brothels or street pimps for sexual exploitation". Thousands of children as young as 12 work as prostitutes in countries including Taiwan, Venezuela, Dominican Republic, Peru, Brazil, Canada and the United States (52).

In the United States, a typical case history involved a 13 year old girl in New York City (53).

SPECIES IN TRANSITION

Gerd Theissen, a German scripture scholar, was quoted in 1996 as saying that the *missing-link* between *apes* and *true-humanity* had been found and that it was "us." He said that we were in a *stage* toward true humanity and that we had not reached it (54).

Human history, as written and through archaeology, clearly indicates that the human species has not changed *physically* in some 100,000 years. Only *cultural* evolution has occurred. This has been a learning process. This means that some primitive instincts continue to dominate human behavior. *Cultural* evolution must include all of the *creative* and *humanizing* effects of *rational*

– and enforceable – laws and scientific discoveries which can extend the quality of life for all world citizens.

This strangest *human* creature includes the most cruel and destructive as well as the most creative and caring of all the animals on the planet Earth. The majority of Earth's *six billion* citizens struggle to survive and to reproduce.

It also includes vast members of the planet's six billion or more people who are ignorant of the most basic facts of history and the *real* world as discovered through the proven methods of science. Billions have been deceived from birth by myths and outright delusions regarding the origins of mankind, our planet and the universe. Some continue to advocate death to *infidels* (non-believers).

In any foreseeable future, the human population will continue to contain many persons who can only be described as "monsters." Some few will continue to hold power which allows them to literally destroy thousands or even millions of helpless or clueless world citizens. (Of course, some *small-time* monsters may live next door as well as across the world).

Some seem born to this monstrous behavior through an inherent *overpowering greed*, probably as a genetic twisted survival instinct, aided by early conditioning . The latter has been called "brainwashing." These factors are the main source of so-called *criminals* who lack the capacity to identify with their victims. This capacity is called *empathy*, not the same as *pity* or *sympathy*. Empathy may or may not possibly be taught (55)(56).

The difference between *education* and *brainwashing* has been concisely defined. Education exposes one to the facts, which are objective. It also exposes one to differing opinions, which are subjective. By contrast, brainwashing is a concentrated effort to erase ("wash the brain of") someone's opinions(usually about politics or religion)and replace them with different *beliefs*. The latter may be forcible, but brainwashing also may occur simply through intensive persuasion and repetition. "A washed brain is soon filled with ideas — someone else's (57)."

Overwhelming *greed* can only be counteracted through a vast increase in *self-knowledge* on a global level, together with *humane* and enforceable laws. Again, the most basic and primitive instincts are for *personal-survival* (causing the almost universal denial of the permanency of an individual's death) and *species-survival* (the chief drive behind sexual reproduction).

Because DNA-based personality traits are unchangeable, the only hope for controlling the most dangerous persons is through the *civilizing* effects of education. Not to include indoctrination into some fanatical religious dogma, where doubters are subjected to humiliation, expulsion, torture or execution — as documented over thousands of years of human history.

The possible beneficial effect on a future for mankind, provided by some major world religions, may continue to depend upon certain basic principles. These have been described in convincing detail by a very competent scriptural historian, teacher and author (58).

Thus, the *towering evils,* most threatening to human survival, are seen as *greed* and *ignorance.* Ignorance can be seen as the greatest danger, if only because it is treatable. Greed is part of human nature, although it might be understood as a kind of perversion of the ancient and primitive *survival instinct.* It surely cannot change without another million years of *physical evolution* (59).

This topic of *Human Nature* deserves further consideration of the frequent questions regarding tragic events, whether caused by *incompetence* or by *conspiracy* — or by a *combination*. This topic will consider the differences between *fools* (persons *fooled* or *deceived*) and *criminals* in power – with examples of *monsters* or *madmen* in history.

Criminals are again defined as persons who knowingly and deliberately cause great harm to others, usually innocent victims.

Fools, again, are here defined as persons who have been *fooled* or deceived. They may include persons who have had no access to reliable information about a situation or important facts of human history. This definition is not meant to include persons referred to as *developmentally disabled* — formerly referred to as mentally retarded — nor to persons who have recognized biological brain disorder manifested by grossly apparent delusions and/or hallucinations.

INCOMPETENCE OR CONSPIRACY

Conspiracy is often the trademark of criminals, acting together to harm others — and not in self-defense, but usually motivated by greed. This has been the true motivation for most of the never-ending human wars.

Criminals often use *fools* (ignorant or deceived persons) to conceal their own involvement. The victims are always found on both sides of any war. For several hundred years, the young men who were betrayed into killing each other were referred to as *cannon-fodder*.

The absolute winners in any war are those who make the weapons and loan the money. Of course, past history has shown *rewards* for the "winning-warriors" to consist of pillage, rape and the taking of slaves. In addition, those who have promoted some wars have done so in the name of some god or of religious necessity — such as saving souls or destroying non-believers.

A basic rule for avoiding public disasters is to beware of all leaders, whether self-appointed or elected. All possible information about them should be made available and every responsibly citizen should vote accordingly. Never should anyone fear to ask "why" or be unable to "follow the money trail." In Part Four, Self-Directed Evolution, this subject will be discussed in greater detail.

HISTORIC MONSTERS AND MADMEN DEFINED

The first need here is to define *monsters and madmen.* The term *monster* describes anyone who deliberately tortures and/or kills *innocent* people, without any evidence of compassion or empathy. This is surely the most heinous of all human crimes.

The term *madman* is to describe a human who has obvious delusions of either grandiosity or paranoia. These have often been part of the personality of *monsters.* Some historic examples will be considered now in this writing.

The *instinctive nature* of those persons who have launched wars of conquest has not changed in the human species' 100,000 years history. Greed and arrogance and cruelty typify the most dangerous aspects of inherent *Human Nature.*

In today's world, the future of mankind may be in much greater danger than in recent past centuries, if only due to the invention of *weapons of mass destruction* and the potential for their prompt delivery anywhere in the world.

As part of understanding *human nature,* it seems appropriate to now briefly examine a few persons who have been very responsible for commission of the most horrible *crimes-against-humanity.*

My reader is encouraged to pursue their own research later. The *Search* button on the Internet is an astonishing source.

Only a few of the most notorious persons in history who surely qualify for the designation of *monster* or *madman* are included in the following short list:

Genghis Khan	(1164-1227 AD/CE)
Tamerlane	(1336-1405 AD/CE)
Cristoforo Colombo	(1451-1506 AD/CE)
Hernando Cortes	1485-1547 AD/CE)
King Henry the Eighth	(1509-1547 AD/CE)
Napoleon I	(1789-1821 AD/CE)
Joseph Stalin	(1879-1953 AD/CE)

Adolf Hitler (1889-1945 AD/CE)
Saddam Hussein (1937-2006 AD/CE)

Note the extensive *omissions* in the above list, including some very prominent religious leaders in history, e.g., Pope Gregory IX. He initiated the Inquisitions in 1231 AD/CE. (See Chapter Four: *Catholic Inquisitions.*)

The following brief biographies may provide some insight into what motivated three notorious mass murderers during the last century. They include Joseph Stalin, Adolf Hitler and Saddam Hussein.

JOSEPH STALIN (1878-1953) became the absolute dictator of the Soviet Union after 1928. He was born 1878 in Gori, Georgia, Russian Empire. Despite some uncertainty regarding his real father, his legal father, Vissarion Dzhugashvili, drank and often beat his wife and son, Iosif (Joseph). One of his childhood friends believed this made Iosif "as hard and heartless as his father."

Attending the Gori Church School, Iosif was forced to speak Russian and was mocked for his Georgian accent. At age 14, he graduated first in his class and then attended seminary on a scholarship. A gifted singer, he had an income from singing at weddings. He was known as a poet before becoming a *revolutionary.*

In seminary, he was reading Karl Marx instead of the Bible. In 1899, he and several classmates were expelled for their *revolutionary* activities. Some additional insight might result from the three times he was sent into "penal exile" (1902,1908 and 1913). Although he escaped twice before, he remained 4 years in a "small hamlet on the Yenisei River." While there, he fathered a son by a 13 year old orphan girl named Daria (60).

More details regarding his childhood and as a *Marxist,* his three episodes of exile as a criminal, and his rise to power (1917-1927)are also available at the above Internet address.

In the 1930s, during his industrialization and "The Great

Purge", he was considered responsible for the deaths of millions of people.

Under his rule, the USSR did play a major role in defeating Nazi Germany and in the USSR becoming one of the two *superpowers* in the world (61).

ADOLF HITLER (1889-1945) was born in Braunau am Inn, Austria-Hungary. He was selected for further consideration because he was so involved in the most barbarous war in history. This has been called the *Holocaust* (1933—1945), with the annihilation of some six million Jews. Another 9-10 million persons murdered included gypsies, Poles, Ukrainians and Belarussians and disabled persons (62).

Information regarding Hitler's family, childhood and World War I experiences may shed more light on his character and motivations (63).

His father, Alois Hitler (1837-1907) was a customs official and his mother, Klara Polzl (1860-1907) was Alois' third wife.

Alois was illegitimate and kept his mother's last name, Schicklgruber until 1876 when he took the name of his step-father, Johann Georg Heidler. It was rumored that Alois mother, Maria, had become pregnant while working as a servant in a Jewish household.

As a child, Adolf was whipped by his father, Alois. He resolved to never cry again. He was a good student until he failed his first year of high school and had to repeat it. Adolf had conflicts with some Jewish fellow students and this was believed to have started his anti-Semite fixation.

In 1905 at age 16 he dropped out of high school and lived a *bohemian* life in Vienna. His passion for painting and architecture resulted only in repeated rejections. Out of money, he was in a homeless shelter and then, by 1910, in a house for poor working men.

In Vienna, he may have been influenced by a *hotbed* of religious prejudice and 19th century racism. Later, in his book, *Mein Kampf,* he referred to Martin Luther as a "great warrior, statesman and reformer", compared to Wagner and Frederick the Great. It seemed significant that *Kristallnacht* began on Luther's birthday.

In May 1913, he received money from his father's estate and moved to Munich, partly to avoid conscription into the Austrian army. He was arrested and returned to Vienna, but found physically unfit, was allowed to return to Munich.

In August 1914, Adolf enlisted in the Bavarian Army and then served in France. He was decorated for bravery twice – in 1914 and 1918. In July 1919, he was appointed as a *spy* for the Reichswehr, looking for communists, Jews and anyone connected with the Weimar party.

Discharged from the Army in March 1920, he returned to Munich and began working fulltime for the German Workers' Party (DAP). By early 1921 he was speaking to large crowds – nearly 6,000 once in Munich.

In July 1921, he demanded dictatorial powers and was given the title of Fuhrer of the National Socialist (Nazi) Party.

In November 1923, after a confrontation with his rival, Gustav von Kahr, he and his followers stormed the Bavarian War Ministry to overthrow the Bavarian government – with no success. In a confrontation with police, 16 of his party members were killed.

He was soon arrested, charged and convicted of high treason. On April 1, 1924, he was sentenced to 5 years in Landsberg Prison. He had much fan mail and received favorable attention from the guards. In December 1924 he was pardoned and released from jail – but briefly banned from public speeches.

While in prison he dictated *Mein Kampf (My Struggle)* and dedicated it to the Thule Society. The book, in two volumes, sold 240,000 copies between 1925 and 1926. By the end of WW II about

10 million copies were sold or given away.

It is interesting that his preoccupation with syphilis took up 14 pages in his book. He called it a "Jewish disease."

His appeal to Germans was in awakening "an offended national pride", and bitter resentment over the Treaty of Versailles with loss of industrial land, colonies and ridiculous *reparations.* He blamed "international Jewry" and the failed Weimar Republic.

After a "bitter ten-year political struggle" he came to power. In 1933, the ailing President von Hindenburg made Hitler the chancellor. Hitler promised "a Greater Germany", abrogation of the Treaty of Versailles, restoration of Germany's lost colonies, and the destruction of the Jews. In 1934 he began full-scale rearmament.

In 1934, after the death of President Paul von Hindenburg, he declared himself Fuhrer, combining his role of Chancellor and, now, President.

In 1935, the Nuremberg Laws deprived Jews of their German citizenship and forbade their marriage to non-Jews. That same year, he withdrew Germany from the League of Nations. In 1936 he reoccupied the Rhineland.

In November 1938, the Holocaust may have begun with *Kristallnacht* (the night of broken glass)– on the birthday of Martin Luther (1483-1546).

Also, in 1938, Austria was annexed and Czechoslovakia dismembered in March. He invaded Poland in September 1939, precipitating WW II.

His continued rise to power as the mesmerizing leader of the Nazi Party (and Germany's *savior,* resulted in the rapid placement of Germany on a war footing, the *Holocaust,* the conquest of Europe by *Blitzkrieg* and failure in Russia. This can be found in several references. These include his death by suicide in Berlin as the Red Army arrived (63)(64)(65)(66) .

The *Holocaust* (1938-1945) resulted in the deaths of some six million Jews – about 2/3 of the pre-war European Jewish population. In addition to the Jews, some 9-10 million others were annihilated – Gypsies, Poles, Ukrainians, homosexuals and disabled persons (67)(68)(69).

At the Nuremberg War Crimes trial (1945-1946), convened by the United States, Great Britain, France and the USSR, death sentences resulted for Hermann Goering, Joachim von Ribbentrop, and Julius Streicher.

Significantly, Hitler's adoption of the *Swastika,* an ancient sun symbol found in many societies and religions, seems related to his grandiose and probably delusional ideas of "saving Germany." This may qualify him for the term *madman* as well as *monster.*

SADDAM HUSEIN (1937-2006)was born in a small village of al-Auja, near Takrit, Iraq. Most of his childhood was in a mud hut in mostly Sunni Iraq. His father, Hussein al-Majid, either died or abandoned his family. Saddam was raised alone by his mother until she married Ibrahim Hassan, a sheepherder, "brutal and a thief". He made Saddam steal chickens and sheep to sell. At age ten, Saddam moved in with his maternal uncle, Khayrallah Tulfah in Bagdad. This uncle, with pro-Nazi leanings, transferred his anti-British attitude to Saddam.

At the age of 16 he failed to be admitted to the prestigious Bagdad Military Academy due to his poor grades.

In 1956 he participated in a failed coup against King Faisal II. In 1957 Saddam joined the Baathist party. In 1958 a group of non-Baathist army officers succeeded in overthrowing the king. In 1959 Saddam and other Baathist supporters failed in an attempt to assassinate General Abdul Qassim – because he had overthrown the king ! Saddam fled to Syria, then to Cairo, where he spent 4 years.

In Egypt, at age 24, he completed high school. About that time he was arrested twice for chasing a fellow student with a knife –

over political differences. In 1961, he entered the Cairo University School of Law but did not graduate.

In 1963, Baathist army officers tortured and killed General Qassim – and mutilated many of the general's followers.

Saddam rushed back to Iraq to join the "revolution." He soon became an interrogator and torturer at the infamous "Palace of the End." He quickly rose through the ranks due to his extreme efficiency as a torturer.

In 1964, he was arrested for plotting the overthrow of president Abdul Salem Aref.

After 1966, when Saddam escaped from prison, and after the 1968 Revolution, his rise to power continued – for the next 10 years. In January 1976, he attained the rank of General, and in 1979, President of Iraq.

He and his family began to take over the country's oil and industrial enterprises. With the help of several assassinations, he took control of many leading businesses. In 1979, to *cleanse* the Baathist party, he had 20 members *systematically* killed – as non-party-faithful.

Saddam had long believed the Kurds and Shiite Muslims were a serious threat. In 1978 he had the Iraq government issue a memorandum stating that anyone whose ideas were in conflict with the leaders of the Sunni dominated Baath party would be subject to *summary execution.*

In 1980, Iraq invaded Iran, also with a Shiite majority. This 8-year war was because Saddam considered Iranian Shiites a threat to his Sunni minority power base.

In July 1982, a failed assassination attempt on him in Dujail was followed by mass killings of 148 Shiites in that village.

In 1983, Masoud Barzani, leader of a Kurdish group fighting the Baath oppression, sided with the Iranians in the war. Saddam had some 8,000 members of Barzani's clan abducted. These included hundreds of women and children. It was assumed they were slaughtered. Thousands were discovered in mass graves.

In March 1988, during the war, Iran invaded North Iraq with Kurd assistance, capturing Halabja. The next day, the Iraqi air force bombed Halabja with poison gas, causing some 5,000 civilian deaths – including children. This was for participating with Iranians in the "attempted overthrow of his country."

A list of Saddam's genocidal crimes were between 1986-1989. They were called the al-Anfal Campaign. He ordered the slaughter of every living human – and animal – in parts of the Kurdish North. An estimated 182,000 men, women and children were killed. (The 1988 poison gas attack on Halabja killed *only* some 5,000 people.)

He blamed the attacks on Iran. The U.S. president, Ronald Reagan, supported this *cover story.*(Perhaps because Ayatolah Khomeini was in Iran).

Another genocidal campaign involved the Shiite *Marsh Arabs* in Southeastern Iraq. By destroying most of the regions marshes, he destroyed their food supply. Some starved and some migrated elsewhere. His motivations are unclear.

The (first)Persian Gulf War really began when Iraq invaded Kuwait on August 2,1990 – supposedly over oil prices and control of the Persian Gulf.(There are accusations of the U.S. deceiving him into thinking they would not interfere !)

Hussein proceeded on August 2nd to *annex* Kuwait as the 19th province of Iraq. The U.S. president, George H.W. Bush, feared that Saddam planned to take over the whole regions oil supplies. The UN Security Council authorized economic sanctions against Iraq. At the request of the Saudis, Bush sent 230,000 troops to protect Arabia. *Operation Desert Shield* began August 6, 1990. Due to a large buildup of Iraqi forces, U.S president Bush ordered another 200,000 troops to prepare for offense action.

On January 16, 1991, the U.S. Congress approved "the most devastating air assault in history" against military targets in Kuwait and Iraq. On February 24th the U.S. president ordered a ground war to begin. In 4 days the Iraqi forces were over-whelmed. The Iraqi army then set fire to over 500 Kuwait oil wells. On March 3rd, Iraq agreed to abide by all UN Resolutions.

A cease-fire occurred on April 3,1991.There were 532,000 U.S. military, with only 147 battle deaths (70).

The author of this book that you are reading believes that vital information about the Middle East has not been generally available to U.S citizens. It was not known to him until his recent research for this book. This will be presented next.

One source describes the aftermath in Iraq of the first Gulf War in 1991. In great detail, it describes the massive destruction of the country's infrastructure, with over $ 200 billion in property damage, as many as 100,000 deaths and five million displaced persons (71).

The second source – which every U.S. citizen should carefully study – is a lengthy treatise by 23 recognized authorities on the Middle East. In 569 pages, it describes the 1991 role of the United States in the first Gulf War – in terms of the *New World Order* (72).

Prior to 1991, Iraq was the world's second largest oil producer and was said to have had the highest standard of living in the Middle East. It had one of the most powerful armed forces in the Arab world, a "complex" health care system, giant hospitals, and an extensive school and university system.

The United Nations sanctions against Iraq in August 1990 were among the most severe in the history of the UN. According to one source, the sanctions prevented the country from rebuilding its destroyed infra-structure and making needed repairs. The Iraq economy collapsed, the oil industry was headed for demise, half its 3 million date trees were dead, the most severe drought in decades occurred in the Winter 1998-1999.

The UN Development Program estimated a cost of $7 billion was needed to restore electricity capacity to its 1990 capacity.

Available potable drinking water was far below the 1991 level before the war. The UNICF estimated that 500,000 children under the age of 5 years had died between 1991 and 1998 due to the war and the UN sanctions. Maternal mortality had become the leading cause of death among were no longer attending school – for economic reasons (73).

At this point, there is surely more than enough personal information about Saddam Hussein to consider him to be one of the great *Monsters* in recorder history. Many of his atrocities are not described in the above text, but there was certainly evidence of his paranoia. There is evidence until 1991, that many of his crimes-against-humanity seemed to have occurred with the full knowledge of the U.S. government – which decided to support him in the war against Iran (74). (It could be assumed this was due to the behavior of the Ayatollah Khomeini in Iran.)

In March 2003, the U.S. and other coalition forces invaded Iraq. Some of the *crimes-against-humanity*, which Saddam Hussein and seven others were accused of at his war-crimes trial in Baghdad are available (75)(76).

It seems of importance to be aware that two of the Iraqi attorneys appointed for the defense of Saddam Hussein were apparently assassinated – on October 20,2005 and November 8,2005. Another attorney fled the country after being wounded.

On June 27,2006, two of Saddam's voluntary American defense attorneys, Ramsey Clark and Curtis Doebbler, in Washington, DC, provided a documented statement alleging that the trial was unfair and the U.S. failed to protect the defense attorneys, "deliberately ensuring an unfair trial."

At this point there may be enough information for my reader to have some insight into the mind of a *monster.* Further detailed historical information about this man until his execution in Baghdad as a war criminal at the end of 2006 is available (77).

In addition to *monsters and madmen,* there seems to be another category which is difficult to classify – *fools in power.* Too often they are there only to carry out the agenda of some obscure person or persons unknown. [Consider Bob Woodward's book titled *Bush at War.* New York. Simon and Schuster, 2002]

FOOLS IN POWER AND HIDDEN AGENDAS

Even today it may shock some U.S. citizens to learn that their own government has been secretly involved in major crimes over a period of years – crimes that harmed U.S. citizens – and the image of their country. Some presidents have been promoted and elected because they would follow the directions of some hidden power-brokers. Officials appointed to high offices have been chosen in the same manner. There is considerable evidence for this in a number of publications – surely not seen by most voting citizens.

When discussing fools-in-power, this present writer believes it is most relevant to first consider the U.S. presidents, competent and incompetent. Some few may have deliberately and knowingly caused great harm to the many – for personal gain (78)(79).

Of most current interest could be the recent presidents from Ronald Reagan to George H. W. Bush to William Clinton to George W. Bush. Let the reader judge.

This author's second priority is *hidden-agendas.* These have often been associated with the highest elected or appointed officials (80)(81)(82)(83)(84) .

The conclusion by the present author, based on a study of the above references – and other sources – is that the U.S. voters have repeatedly been deceived by elected officials.

These persons were probably aware of their wrongdoing, but were responding to powerful groups who had arranged their election.

The ordinary tax-paying hourly-wage-slave and the *voluntary* military sent to invade other countries suffer the most – second only to the massive civilian casualties. As in all wars, the arms makers and money-lenders amass great wealth.

In World War II, the U.S. was attacked. In each war since, the U.S. has invaded another country: Korea, Viet Nam, and Iraq twice.

Most informative and important today is a detailed account of *terrorist* attacks during the George W. Bush administration – beginning with September 11, 2001 (85)(86)(87).

REFERENCES AND NOTATIONS: CHAPTER TWO

HUMAN NATURE OVERVIEW

1) *Extinctions Past and Present.* See www.worldbook.com/fun/wbla/earth/html/ed12.htm [NOTATION: Of some 5 major extinctions of life on Earth, the largest took place about 240 million years ago at the end of the Paleozoic Era. At that time most species were in the oceans. The most recent was about 65 million years ago when a large asteroid hit our planet. There are many theories as to causes of these extinctions. Scientists agree that each was related to major climate changes, regardless of the causes. The same is true of many species extinctions occurring today. *We live on a very fragile planet* ! *We must quickly understand our responsibility for our continued survival as a species.*]

PERSONAL SURVIVAL INSTINCTS

LIFE-AFTER-DEATH MYTH

2) Dawkins, Richard. *The God Delusion.* New York. Houghton Mifflin, 2006.

HISTORIC REVOLUTIONS

3) www.en.wikipedia.org/wiki/Revolution
4) Mulroy,Kevin (Editor in Chief).Almanac of World History.
5) Stearns,Peter(General Editor).Encyclopedia of World History – 6th Edition. New York and Boston. Houghton Mifflin, 2001. Species Survival (sexual) Instincts

SPECIES-SURVIVAL (SEXUAL) INSTINCTS

SEXUAL BEHAVIORS IN HISTORY

6) Taylor,Timothy. *The Prehistory of Sex: Four Million Years of Human Sexual Culture.* New York and London. Bantam Books,1996. [NOTATION: Timothy Taylor was an archaeologist at the University of Bradford, U.K.(The title referring to *Four Million Years of Human Culture* apparently included the *hominid* – or *ape* ancestor – of both humans and chimpanzees.) See Chapter One: The Evolution of Human Sexual Culture: pages 19-51.[Page 8 regarding rock art and female goddess figurines dating from the most recent Ice-age. Also see pages 172-175 with rock engravings of men having sex with animals. INDEX topics include: adultery; animal sexual behaviors, ancient Egypt (incest, contraception, male masturbation, homosexual monks, brutal rapes (2000 years ago),Roman brothels, pre-human sex, sex slavery and prostitution, rape, abortion.]

7) Wendt,Herbert. *The Sex Life of Animals.* Translated from a 1962 German publication by Richard and Clara Winston. New York. Simon and Schuster, 1965. [Explicit prose and graphic illustrations.[NOTATION: *Ancient Beliefs,* reasonable and unreasonable, are examined. See pages 300-301: "Animals have no objection to copulating with members of another species if normal mates are lacking. Compare the historic record of hanging humans for having sexual contact with animals, as well as killing the animal involved. Among humans, "The *abnormal* (not related to preservation of the species) can survive alongside the *normal*". This statement referred to masturbation, homo- sexuality , gratification from "all sorts of physical contacts .. or other inappropriate objects."]

8) Chauveau,Michel. (Translated from the French. *Egypt in the Time of Cleopatra.* Itheca and London. Cornell University Press,2000.[See index items: Royal incest, divorce, legal autonomy of women]

9) Dersin,Denise (Editor).*What Life was Like on the Banks of the Nile: Egypt 3050-30 BC/BCE.* Time-Life Book. [See index items regarding equality of women, divorce, *Royal* incest and harems(See pages 63,80,94). Beautiful illustrations]

10) White,J.E.Manchip. *Ancient Egypt: Its Culture and History.* Mineola, NY. Dover Publications,1952,1970.[See Index items: Marriage, Royal incest marriages(to protect the sacred bloodline),polygamy, harems, concubines, infidelity of Egyptian women, slaves, erotic pictures, brothels, cult of Min(sacred concubines and sacred prostitutes),women – first child when age 12,followed by 6-7 more.]

11) Dersin,Denise (Editor).*What Life was Like at the Dawn of Democracy: Classical Athens (*Between 525-322 BC/BCE). [See same items as in Ref 9).

12) Dersin,Denise (Editor).*What Life was Like When Rome Ruled the World: The Roman Empire* (100 BC/BCE-200 AD/CE).[See index for these items: adultery, marriage, divorce, army wives, slaves, prostitutes, weddings, Vestal Virgins. Beautiful illustrations.]

13) Clarke,John R. *Roman Sex:100 BC/BCE – 250 AD/CE.* New York. Harry Abrams, Inc.,2003. [See *Conclusion: Sex Before Puritan Guilt.* Many historical illustrations.

14) Davies,John. *The Celts.* New York. Sterling Publishing, 2000.[See index item: Human sacrifice (pages 74,84)]

15) Cherici,Peter. *Celtic Sexuality.* London. Gerald Duckworth, 1994.[Years covered c.400-1600 AD/CE. Seen on a TV Series,2000-2001. Many illustrations. See Index:adultery, brothels, castration, celibacy, chastity, childbearing, homo-sexuality, incest, love, marriage, masturbation, morality, original sin, polygamy and polyandry, prostitution, rape, vir-

ginity, Roman sexual customs and Christianity versus Celtic paganism. See Chapter 2: The Land of Innocence.]

16) Dersin,Denise (Editor).*What Life was Like Among Druids and High Kings: Celtic Ireland*(400-1200 AD/CE. Alexandria,VA. Time-Life Books, 1998. [See Index items: childbearing, druids, legal status of women, marriage, monks and nuns, slaves, childbirth.]

17) Caner and Caner. *Unveiling Islam.* Grand Rapids, MI.Kregel Publications, 2002.[NOTATION: Muhammad ibn Abdallah lived between *570-632* AD/CE. See Chapter 8: Women, Love, Marriage and Property. *Excerpts*:"Islam teaches that women are inherently inferior to men." "Muslim men are allowed to marry 2,3 or 4 wives." "Muhammad received special dispensation directly from Allah to marry as many as he wished." In Chapter 2, his eleven wives and two concubines are listed. In about 623 AD/CE, his friend, Abu Bakr As Siddiq, gave Muhammad his six year old daughter, Aishah, to be another of his wives. Sexual relations began only when she was age 9. He died with his head in her lap. Muhammad gave his own 12 year old daughter, Fatima, to his cousin, Ali bin Abu Taleb]

18) Golden,Arthur. *Memoirs of a Geisha*.New York.Random House/ Vintage Books,1997. [NOTATION: This historical novel has its setting in Japan during the worldwide *Great Depression* of the 1930s. A quaint, but very apt, description by the Geisha was that men had *restless eels,* always in search of as many as possible (female)*caves* to explore.

EMANCIPATION OF WOMEN

ADOLESCENT SEXUAL BEHAVIORS

19) http://en.wikipedia.org/wiki/Adolescent_sexual_behavior Legal Age for Consent [Text p.41]

20) http://www.avert.org/aofconsent.htm [NOTATION: Worldwide Age of Consent data]

LEGAL AGE FOR CONSENT

SELF-EXTINCTION BEHAVIORS

ENDLESS WARS AND ASSOCIATED ATROCITIES

21) http://en.wikipedia.org/wiki/Atrocities

22) LeBlanc,Steven. *Constant Battles.* New York. St. Martin's Press,2003. See pages 8,81-82.

23) Zinn,Howard. *Terrorism and War.* New York and London. Seven Stories Press,2002.
24) Carr,Caleb. *The Lessons of Terror:The History of Warfare Against Civilians – Why it Has Always failed and Why it Will Fail Again.*
25) Gutman,Roy;Rieff,David (Editors).*Crimes of War: What the Public Should Know.* New York. W.W. Norton, 1999.
26) Meron,Theodor.*War Crimes Come of Age: Essays.* New York. Oxford University Press,1998.

CRIMES AGAINST HUMANITY

27) www.who.int/mediacentre/factsheets/fs241/en . . . [NOTATION: *Female genital mutilation*(FGM). World Health Organization Media Centre. This ancient custom continues mostly in 28 African countries, but also in the Middle East, Asia, Europe, Australia, Canada and the U.S. Other than Africa, the other countries are represented by immigrants. Some Muslim communities continue the practice, believing it is required by their faith. The practice long predated Islam. This monstrous physica assault on innocent females amounts to torture and a crime-against-humanity.It demands laws, which are enforceable on a global scale.
28) www.dominionpaper.ca/accounts/2005/02/11ending_fem.html Richardson,Gemma. Ending Female Genital Mutilation (FGM) ? [NOTATION: This article on the Internet pertains particularly to Kenya. It discusses prevalence in other African countries (with map).]

GENOCIDE AS A CRIME-AGAINST-HUMANITY

29) www.historyplace.com/worldhistory/genocide/holocaust.htm
30) War Crimes. Columbia Encyclopedia – Sixth Edition,2007. Also see http://www.encyclopedia.com/printable.aspx?id=1E1:warcrime

GENOCIDES IN HISTORY

31) War Crimes. Columbia Encyclopedia – Sixth Edition,2007. http://www.encyclopedia.com/printable.aspx?id=1E1:warcrime
32) Gutman,Roy;Rieff,David. *Crimes of War.* New York and London.W.W. Norton,1999. See pages 58-65 re Cambodia.
33) Meron,Theodor.*War Crimes Law Comes of Age.* Oxford. Clarendon Press, 1998. See page 296 re Cambodia.
34) www.historyplace.com/worldhistory/genocide/bosnia.htm
35) Gutman,Roy;Rieff,David. *Crimes of War.* New York and London.W.W. Norton,1999. See pages 312-318.*Rwanda genocide.*
36) http://en.wikipedia.org/wiki/Second_Congo_War.

37) http://worldnews.about.com/od/sudan/a/darfur_peace.htm?p=1 Massacres in History [text p.44]
38) Massacres are listed at: http://en.wikipedia.org/wiki/List_of_massacres

TORTURE AND RAPE

39) Ayton-Shenker,Diana; Tessitore,John. *A Global Agenda: Issues Before the 56th General Assembly of the UnitedNations(2001-2002.* See *the International Criminal Tribunal for the Former Yugoslavia,* pages 248-260.
40) Swain,John. *The Pleasures of the Torture Chamber.* New York. Dorset Press,1931 and 1995. Especially note Chapter IX:*The Holy Inquisition.*
41) Cheney, Malcomb. *Chronicles of the Damned: The Crimes and Punishments of the Condemned Felons of Newgate Gaol* Dorset, UK. Marston House,1992.
42) Rejali,Darius. *Scholar studies torture, ancient and modern.*Portland ,Oregon.The Oregonian,10/06/03. [NOTATION: On that date, this eminebt scholar, Rejali, had three books published or pending.]
43) Gutman,Roy;Rieff,David (Editors).*Crimes of War.* New York and London. W.W. Norton, 1999. See pages 323-329: Rape .
44) Ayton-Shenker,Diana;Tessitore,John.*A Globa Agenda:Issues Before the 56th General Assembly of the United Nations.* [NOTATION:See *The International tribunal for the Former Yugoslavia, pages 248-260.*]
45) Chang,Iris.*The Rape of Nanking:The Forgotten Holocaust of World War Two.* New York and London. Penguin Books, 1997.
46) Gilbert,Martin. *The Second World War: A Complete History.* New York. Henry Holt, 2004. [NOTATION: Sir Martin Gilbert was knighted for his original version of this publication in 1999.]
47) Beevor,Antony. *The Fall of Berlin – 1945.*[NOTATION:See Data on rape of German women and girls by the Red Army on pages 28,32,118,188,326,410.]
48) Meron,Theodor. *War Crimes Come of Age: Essays.* New York. Oxford University Press,1998.[NOTATION: A very important *must-read* book. See page 47: "In World War Two, rape was tolerated and horrifying was even used in some instances as an instrument of policy."In occupied Europe,thousands of women were subjected to rape and thousands more were forced to enter brothels for Nazi troops." "Rape of German women by Soviet soldiers appeared tolerated."]
49) Gutman,Roy;Rieff,David. *Crimes of War.* New York and London. W.W. Norton, 1999.[NOTATION: See pages 323-329: "Rape as a spoil of combat." "Rape has been considered a war crime for years." "It falls under

the definition of of torture .. according to the International Criminal Court (ICC)." Crimes-against-humanity also include sexual slavery, enforced prostitution, enforced sterilization – whether during war or peace – and if of a widespread nature."]

50) http://web.amnesty.org/library/Index (surviving rape in Darfur). Human Trafficking [text page 46]

51) http://english.pravda.ru

CHILDREN AS PROSTITUTES

52) Child prostitution becomes a global problem with Russia no exception. Seen at http://english.pravda.ru/print/society/stories/84991. [NOTATION: This lengthy and detailed article is worthy of careful study.]

53) T*he 13 year old prostitute.* New York Magazine. Found at http:// nymag.com/news/features/30018

SPECIES IN TRANSITION

54) Maguire,D.C. *Earth Ethics: Spring-Summer 1996.* www.Religiousconsultation.org/ population_sustainable_dev.htm

55) Kogler,Hans; Steuber Karsten (Editors). *Empathy and Agency.* Boulder, CO and Oxford, UK. Westview Press, 2000.

56) Schelller,Samuel. *Human Morality.* New York and Oxford. Oxford University Press,1992. [NOTATION: This is a detailed and scholarly examination of self-interest versus social conditioning and education.]

57) vos Savant,Marilyn.*Ask Marilyn.*PARADE Magazine, June 6, 2004.

58) Kimball,Charles.*When Religion Becomes Evil: Five Warning Signs.* HarperSanFrancisco, 2002.

59) Minard,Eugene. *Evolution of Gods: An Alternative Future for Mankind.* Portland, OR. Metropolitan Press, 1987.

HISTORIC MONSTERS AND MADMEN

JOSEPH STALIN

60) http://en.wikipedia.org/wki/Joseph_Stalin

61) TIME Almanac – 2006. [Pages 297,679-681, 778]

ADOLF HITLER

62) TIME Almanac-2006. *Headline History: The Holocaust.*

63) http;//en.wikipedia.org/wiki/Adolf_Hitler66) en.wikipedia.org/wiki/Adolf_Hitler

64) http;//en.wikipedia.org/wiki/Adolf_Hitler66)en.wikiped ia.org/wiki/Adolf_Hitler [NOTATION: 50 pages]

65) Evans,Richard J. *The Coming of the Third Reich.* New York and London. The Penguin Press,2004.

66) http;//en.wikipedia.org/wiki/Blitzkrieg

67) Beevor,Antony. *The Fall of Berlin-1945.* New York and London,2002.

68) TIME Almanac-2006.*Headline History: The Holocaust.* [Page 680.]

69) www.historyplace.com/worldhistory/genocide/holocaust.htm

SADDAM HUSSEIN

70) TIME Almanac-2006. *The Persian Gulf War (January-April 1991).* [See page 691]

71) http://www.canesi.org/Enggl/impact.html [NOTATION: This Canadian Internet article was titled Consequences of war bombings and sanctions against the people of Iraq.]

72) Ismael and Jacqueline Tareq (Editors).*The Gulf War and* the New World Order: International Relations of the *Middle East.* Miami and Pensacola. University Press of Florida,1994. [NOTATION: This 569 page dissertation by 23 notable authors is a "must read" for most Americans who may be mislead by the true ambitions of U.S. leaders over a period over many years.]

73) http//www.canesi.org/Enggl/impact.html. [NOTATION: This Internet site was still available at the end of 2007 and the sponsor was named as CANESI, meaning *Canadian Network to End Sanctions on Iraq*]

74) http://civilliberty.about.com

75) http://www.emergency.com/hussein1.htm

76) http://www.rjgeib.com/thoughts/burke/hussein.html

77) http://www.msn.com/id/16389128/

FOOLS IN POWER AND HIDDEN AGENDAS

78) Tolson,Jay. *The Ten Worst Presidents.* U.S. News and World Report, February 26,2007.

79) O'Brien,Cormac. *Secret Lives of Presidents: What Your Teachers Never Told You About the Men in the White House.*Philadelphia. Quirk Books, 2004. [NOTATION: From George Washington to George W. Bush, this a detailed, candid historical account of each man, both good and bad. It has impressive authenticity.]

80) Bowen,Russell S. *The Immaculate Deception: The Bush Crime Family Exposed.* Carson City, NV. America West 66 Publishers,1991.

[NOTATION: This expose is surely a must-read for U.S. citizens. The author's credibility and competence to provide the information in his book seems beyond question. A patriotic teenager, he became a WW II decorated fighter pilot over Germany. After WW II, as a strict and obedient young Catholic, he met the Pope in the Vatican in 1959. Soon after, he was asked by William "Wild Bill" Donovan (a WW I General)to join the OSS (later the CIA) as an *under-cover-agent.* He then spent 47 years as a "mole" or "super-spy" – involving *secret missions.* His first assignment was to Paraguay to study the traffic routes of drugs to the U.S.A. Eventually, he realized that he had been mislead by certain American "leaders" who claimed to be the *guardians* of America, when actually they were then and now – damaging the country. The author is a retired Brigadier General.]

81) Ismael and Jacqueline Tareq (Editors).*The Gulf War and the New World Order:International Relations of the Middle East.* Miami and Pensacola. University Press of Florida,1994. [NOTATION: This 569 page dissertation by 23 notable authors]

82) Brisard,Jean-Charles; Dasquie,Guillaume.(Translated from the French to English by Lucy Rounds).*Forbidden Truth: for Bin Laden.* New York. Thunder's Mouth Press/Nation Books, 2002.

83) Johnson,Chalmers. The Largest Covert Operation in CIA History. Found at http://hnn.us/articles/1491.html [NOTATION: This article begins with "The Central Intelligence Agency has an almost unblemished record of screwing up every secret armed intervention it ever undertook".]

84) Woodward,Bob. *Bush at War.* New York, London, Toronto, Sidney and Singapore. Simon and Schuster,2002.

85) Baker,James;Hamilton,Lee (Co-chairs).*The Iraq Study Group Report.* New York. Vintage Books/Random House,2006.

86) *The 9/11 Commission Report* .

87) Brown,Russell S.*The Immaculate Deception: The Bush Crime Family Exposed.*Carson City, NV. American West. 1991. [ALSO SEE: Abraham,Rick. *The Dirty Truth – The Oil & Chemical Dependency of George W. Bush.* Houston,TX. Mainstream Publishers, 2000.]

PART THREE

Human self-extinction

CHAPTER THREE

THE SEVEN MAJOR TIME-BOMBS

THE SEVEN TIME-BOMBS

(REFS 1-66)

This Chapter is a comprehensive examination of the major forces in the world today which may determine whether the human species will survive or become extinct. The alternative, of course, may only be "the end of civilization as we know it (1) ."

Again my readers are urged to use the materials provided here as *evidence*, to be evaluated (in terms of their *probability)*, and to encourage their own further research. No species has had the power of self-extinction except for humans.

This has become more obvious since the Industrial Revolution, beginning in the mid-1700s. It has led to an astonishing increase in scientific discovery, invention, rapid population increase, massive toxic pollution, and increasing depletion of natural resources. It has provided weapons of mass destruction and the ability to deliver them in hours to any part of the world.

The increasing possibility of human self-extinction includes the following global *time- bombs*:

1) Human population explosion – the *greatest* underlying threat, creating most of the following problems.

2) Environmental toxic pollution, chemical and radiation .
3) Emerging global disease pandemics, e.g., new incurable prion and viral infections.
4) Exhaustion of natural resources.
5) Global warming with devastating climate changes.
6) Nuclear holocaust, possibly as part of the holy wars.
7) Natural disasters, such as massive earthquakes, meteor strikes, and reversal of the Earth's polar magnetic poles.

A combination of several of these at the same time could mean the end of "civilization as we know it (1)."

HUMAN POPULATION EXPLOSION

[REFS 2-11]

The global population explosion could be the single most ominous event threatening the survival of the human species. This could be considered to be more menacing than the continuing danger of a nuclear holocaust, whether related to *holy wars*" or to empire building by international corporations and banks. *Colonization* invites native rebellions. Consider the very long history of the Middle East.

Unrestrained human population growth could be compared to bacteria in a culture medium. They perish when the nutrients are exhausted. A similar comparison could be the uncontrolled multiplication of cancer cells, ultimately destroying their host.

The following reference is an early but very concise evaluation of the "population explosion" and "world politics (2)." By the year 2050, the planet's population will probably exceed 9.3 billion people. Half of the increase will occur in six countries: India, China, Pakistan, Nigeria, Bangladesh, and Indonesia. Some countries, especially those that are predominantly Catholic or Muslim, object to any candid discussion of sexuality in that it "promotes abortion." (3).

Following the U.N. 1994 Cairo population conference, regarding the urgent need to attain "zero population growth," there were indignant protests by Muslim scholars and the Vatican. They insisted that the resulting "statement" by the Conference encouraged abortion and promiscuity (4).

In the foreseeable future, a stable world population can only be possibly by drastically reducing the number of births or increasing the number of deaths, probably only through some *natural disaster.* About 30 years ago, the "carrying capacity" of planet Earth was estimated as between 10 and 15 billion humans (5)(6).

A definition for "carrying capacity" and a formula for predicting this may be found in the following references. Carrying capacity has been defined as "the theoretical maximum number of organisms in a population, usually designated by *K*, that can be sustained in a given environment (7)(8).

Again, in the foreseeable future, a stable world population can only be possibly by drastically reducing the number of births or increasing the number of deaths, probably through some natural disaster.

Homo sapiens sapiens barely survived from their beginning as a distinct species around 100,000 years ago (YA). By around 8,000 BC/BCE (10,000 YA),there was an estimated world population of only *8 million people.* This estimate was by anthropologists and historians.

During the 8,000 years to 1 AD/CE, the population increased to only 300 million. With the gradual ending of the last Ice-Age, this increase is believed to be due to the change from a hunter-gatherer culture to agriculture and animal domestication. During those 8,000 years, the annual growth rate (number of births exceeding deaths) has been calculated at about .036 persons per thousand population.

Between 1 AD/CE and 1750 AD/CE, the population increased

to an estimated 800 million. The current tremendous acceleration in the human growth rate began with the "Industrial Revolution" around 1750 AD/CE.

By 1800 AD/CE, there were about 1 billion persons (1000 million). A growth rate of about 5.2 persons per 1,000 population resulted in a world population of around 1.3 billion by 1850 AD/CE.

To reach 1.7 billion in the year 1900 required a birth rate of 5.4. In 1950, the United Nations estimated a world population of 2.5 billion; by 1960 there were 3 billion; 4 billion by 1974; 5 billion in 1987 and 6 billion is 1999. In 2001, the U.S. Census Bureau estimated the world growth rate at 13 (22 births and 9 deaths per 1,000 population).

The U.N. projected a population of 10-11 billion by 2050 AD (9)(10)(11)).

CHEMICAL AND RADIATION POLLUTION

[REFS 12-30]

The accelerating extinction of animal and plant species, as well as the increasing cancer rates, have been related to massive pollution of the planet's air, water, food and land — by humans.

Suspected or proven "carcinogens" were described in detail and at length in a publication over 25 years ago. After extensive research of the literature, it was concluded that "a high percentage of cancers may be preventable through application of reasonable environmental controls." A multitude of cancer-causing agents were identified, including those found in nature, as well as those introduced into the environment by modern industries (12).

Historic poisonings of the environment have included: the "killer smogs" in Donora, Pennsylvania (1948) and London (1952); the pesticide explosion at Monsanto, Nitro, West Virginia (1949); mercury poisoning deaths in Minimata, Japan (1950-1975); Vietnam "Agent Orange" spraying by the United States (1962-1971);

deaths of dioxin poisoning in Sevoso, Italy, at the Icmesa plant explosion (1976); radioactive pollution from the Three Mile Island nuclear power plant in Pennsylvania (1979); toxic gas leak by Union Carbide at Bhopal, India (1984) and global radioactive contamination from the nuclear power plant explosion at Chernobyl, Kiev, April 1986.

An updated report on the continuing extreme hazards from radioactive waste at the Hanford, Washington, former nuclear-weapons facility considered it to be " a million times more lethal" than the leak at the Pennsylvania Three Mile Island nuclear energy facility. The Bush administration had recently persuaded the Senate to place a rider on a defense bill, renaming the extremely lethal nuclear waste as "incidental" and eligible for quick, cheap surface disposal rather than deep-underground sequestration. Water in the Columbia river now risks contamination for thousands of years. Similar problems existed at the Aiken, South Carolina, nuclear complex, near the Savannah River (13)(14).

In April 2007, at the Hanford Nuclear Reservation in the State of Washington, a contractor was responsible for the spill of about 85 gallons of high-level radioactive waste from a tank. [Info source: The Oregonian, 12/02/07]

One of the most frightening documents found by this author concerns water and may be found in the following reference (15)."Chemical contamination of our water supplies has become the most serious environmental problem of our time."

Disasters related to water were initially recognized as being due to the contamination of rivers and lakes and, only later, of underground *aquifiers* (porous layers of sand or rock). The percolation of as little as a gallon of a day of gasoline down through the soil may make the ground-water from wells unfit to drink for a town of 50,000 people.

"Each year in the United States, some 10 billion gallons of sewage, radioactive waste, chemicals and brine are injected deep into the earth." Similar appalling statistics are provided in the reference given above. In addition, it has been clear that the government and industries have been derelict in discovering and remedying the situation. Apparently, only class-action lawsuits have gotten results.

Detailed descriptions of several water related environmental disasters include: Elevated cancer rates in New Orleans, tied in 1974 to some 66 different organic chemicals in drinking water; Atlantic City, New Jersey, which drilled new wells rather than try to clean up toxic chemicals in the "hopelessly polluted" old ones; Fort Edward, New York, about 1982, where well water levels of carcinogenic (TCE) were found to be over 200 times the accepted safe level.

Polybrominated diphenyl ether (PBDE) levels have been increasingly detected throughout the *food-chain* and in mother's milk. These chemicals are used in plastics, TVs, furniture and carpets (16).

The source of the Fort Edwards contaminated water was clearly the nearly General Electric factory, where the TCE was a by-product of electrical capacitor manufacturing. During the course of some 40 years, chemicals had leaked or been spilled, and had leached into the groundwater. Although GE denied responsibility, they paid each homeowner $25,000 to connect with the city water supply.

After World War Two, there was an "explosion" of new chemicals. Synthetics largely replaced cotton, wool and silk. Plastic had replaced wood, glass and metal. "Rubber" tires were now synthetic. In 1985 more than 60,000 chemicals were used in manufacturing with very little known of their effects on health.

Many more examples of water contamination are provided in the reference cited. Additional information concerning the effects of certain chemicals on the nervous system can be found in the following publication (17).

Several lists of dangerous chemicals are available (18)(19)(20).

Probably the most toxic and studied of 75 compounds known as CDDs (*chlorinated dibenzo-p-dioxin*) is 2,3,7,8-TCDD.

Exposure occurs by consuming food containing this chemical. It can cause cancer or a severe skin rash called chloracne. It may be contracted during the manufacture or use of certain pesticides and herbicides. It is formed during the bleaching process at pulp-and-paper-mills and during the chlorination of waste and drinking water at treatment plants. Released into the air, it can be transported even around the planet. It may build up and be concentrated in the food chain, with measurable levels in animals.

More detailed information regarding CDDs is available from the Agency for Toxic Substances and Disease Registry in Atlanta, GA, or at www.atsdr.cdc.gov/facts104.html.

Recently reported research indicates that air pollution from steel mills and major highways contains toxins such as polycyclic aromatic hydrocarbons which produce *heritable mutations* in birds and mice, usually detrimental (21).

An especially competent and comprehensive presentation of research findings in regard to the toxic effects of environmental pollutants, especially chemical, on the human nervous system, has been provided by a multinational group of scientists. This is considered a key reference on this subject (22).

For some 200 years, cancers have been found to be related to environmental causes. Scrotal cancer was found in association with chimney sweeps in London in 1775. "At least three fourths of the human cancers are *environmental* in origin (23)."

Cancers of all types have combined to rank second only to *heart and artery* disease among *deaths* per 10,000 population in the United States: 162.8(1970); 184.3 (1981); 201.6 (1999) (24).

In 2005, the estimated *new* cancer *cases,* of all types, in the U.S. were 1,372,910. The estimated U.S. population in July 2005 was 296,496,649. This results in 376 per 10,000 pop-ulation.

In 2002, all types of cancer *deaths* in the U.S. were 194.0 per 100,000 – or 19.4 per 10,000 population (25).

Deaths from massive chemical poisoning occurred in Minimata, Japan, (mercury), 1950-1975. At Love Canal, Niagara Falls, New York, Hooker Chemical Company had buried some 20 tons of "toxic chemicals," 1942-1953, under what later became a housing development. Uncovered by heavy rains, the waste caused a marked increase in cancers, abortions, birth deformities, epilepsy, and hyperactivity in children.

In 1971 Vietnam, use of "Agent Orange" (mostly dioxin, the defoliant 2,4,5-T) was discontinued by the U.S. military after 9 years of spraying jungles, having resulted in birth defects, deformed babies and a 5-fold increase in a rare form of liver cancer in natives. Who profited from the sale of this chemical?

This particular herbicide – among some 15 types of herbicide used in Viet Nam – was made by Dow Chemical and Monsanato.[Data from Wikipedia website]

RADIATION, both ionizing and non-ionizing, justifies "continuing public health attention as an environmental health hazard." Most research to date indicates no apparent health hazard from normally encountered low frequency radiation, but a connection between high frequency fields and cancer is still under study (26)(27).

Ionizing radiation includes alpha, beta and gamma particles and x-rays. Ionizing radiation is also produced when radioactive materials "decay" as their natural rate, stripping electrons from atoms, creating two charged ions or breaking chemical bonds. Often causing cancer, it includes electromagnetic fields (EMFs) of very high frequencies (VHFs) and short wave length—for example, x-rays, cosmic rays and gamma rays from radioactive materials.

High frequency transmission also occurs with cell phones, microwave ovens and television. These *radio-frequencies* (RFs), produced by alternating currents (Acs), have frequencies between 300 cycles per second (Hertz) and 300 Gigahertz.

Directly-overhead high voltage transmission wires may have very high electrical field strengths which drop off to "normal" at 150-300 feet. Cancer continues to be the primary health concern of EMFs.

Radioactive materials, with long exposure, result in harmful effects, depending upon the amount of radiation. High levels of exposure will cause "radiation sickness," premature aging or death within two months. This, of course, pertains to persons not killed outright.

Exposure to dosages of 50-100 rems results in symptoms which include nausea or vomiting, fatigue, hair loss, diarrhea, bleeding within hours to two months. Dosages of 2,000 rems damage the central nervous system, bring about loss of consciousness and death in hours to days. Such exposure should never occur normally but did with nuclear power plant leaks or "meltdowns," as at Three Mile Island, Pennsylvania (1976) or Chernobyl, Kiev (1986).

Non-ionizing radiation involves low-frequency long wave length electro-magnetic fields as from an electric wall socket. Even when no electrical device is turned on, an electric field remains because the wire remains *charged*.

Extremely-low frequency fields (ELFs) are used in household electrical appliances. Higher frequencies are needed for security systems, computer screens, radio and television, cell phones, microwaves and radar. These electromagnetic fields induce electric currents in the human body and can produce heating or electric shock if the amplitude, frequency and strength are high.

As a non-ionizing form of radiation, ELF has been studied at some length in Great Britain for possible health risks, including cancer. No "convincing" evidence was found for "any effect on biological processes (28)."

In contrast, earlier studies have demonstrated definite adverse effects on animals and humans. In the early 1970s, the U.S. Navy *Project Sanguine* was planned, with a giant antenna in northern Wisconsin, to broadcast ELF signals to circle the globe. After alarming questions were asked, no further public information was made available. When strong protests were lodged by a U.S. Senator, the Navy changed the plans, the location and the name to *Project Seafarer.*

In 1974, a network of 756-kilovolt power lines were planned to link nuclear power reactors in upstate New York and Canada. ELF fields at lower power frequencies had been linked to bone tumors in mice, slowed heartbeat in fish, and multiple physiological changes in rats. The two power companies hired other "experts" and attorneys who attacked the earlier research as the product of "quacks." (29).

An interesting effect of *positive ions* in the atmosphere concerns hot desert winds in the Middle-East and in the South-western U.S. The Israeli wind is called *sharav* and has been linked to elevated serotonin blood levels, headache and "hot temper". This effects up to half the adult population of Jerusalem. Serotonin inhibitors or negative ion generators are believed to help (30).

NANO-TECHNOLOGY, as a growing, largely unknown, possibly very serious global health hazard, is difficult to classify. My reader is urged to consult the Internet.

One excellent recent reference is *A Little Risky Business* in *The Economist.* November 24, 2007.[ALSO see: Seavey, Todd. *Neither Gods Nor Goo: Avoiding both utopian and apocalyptic forecast for nano-technology.* Los Ageles, CA. Reason Foundation, March 2008.][ALSO see: Wang,Zhong Lin. *Self-Powered Nanaotech.*

Scientific American, January 2008. ALSO see: Bergstein,Brian. *Future shock: That shirt may have a power surge.* The Associated Press, February 14, 2008]

GLOBAL DISEASE PANDEMICS

[REFS 31-40]

On a worldwide scale, the major causes of death should be considered in terms of "diseases," with causes and prevention examined. In addition, a newer approach has been made in regard to the impact of disability in the absence of actual death. This has been described in terms of "lost years" due to disability or as "disability-adjusted life-years" (DALYs) (31).

Death *rates* by cause are more useful in comparing different populations at different times to reveal trends.

In the United States in 1999, the most significant causes of death were: heart diseases (266 deaths per 100,000 population); cancer (202); strokes (61); diabetes (25); motor vehicle accidents (16); Alzheimer's disease (16); kidney diseases (13 deaths per 100,000 population) (32).

In the United States, deaths from all types of malignancy (cancer) have steadily increased, from 163 per 100,000 population in 1970, to 184 in 1981 to 202 in 1999, as shown above.

The death rate from diabetes dropped somewhat between 1970 and 1981 (19 to 15) but rose again to 25 per 100,000 (1999). Also, deaths related to kidney diseases rose steadily from 4.4 in 1970 to 7.6 in 1981 and then to 13 per 100,000 in 1999.

A clear trend in evident, *unexplained* , toward increasing deaths due to cancer, chronic kidney disease and, possibly, diabetes. (It should be noted that the main cause of death among U.S. teenagers continues to be automobile accidents.)

In contrast to the U.S. and other "developed" countries, major causes of death in sub-Saharan Africa, Latin America, the

Caribbean, Mid-east and some Asian countries continue to relate largely to infectious diseases, ranging from 10 percent to 13 percent of deaths. "Developed" countries average 6 percent (33).

The *most recent* statistics will be available from the Communicable Disease center on the *Internet.*

Recent data from the World Health Organization indicates that the highest priorities for disease prevention or eradication involve HIV-AIDS (as "a threat to international peace and security"), tuberculosis, malaria and polio. There is increasing concern for the mental health of children, affecting as many as one in five. Large increases have been noted in depression and suicide among children and adolescents (34).

Incurable *prion* (e.g., Mad Cow) and *viral* pandemics (e.g., AIDS) continue to potentially endanger the future of mankind. Prions could potentially become an extreme danger, greater than the AIDS virus (35)(36)(37)(38)(39).

A fairly recent update on "mad-cow disease" or bovine spongiform encephalopathy (BSE) discussed the high cost and probably ineffectiveness of testing all 35 million U.S. cattle annually before slaughter. A better use of these funds to protect consumers would be the inclusion of *all* human cases of suspected *prion* disease for post-mortem examination at the Case Western Reserve University National Prion Disease Surveillance Center. This would greatly assist in the epidemiological search for sources of the disease.

In December 2003, within 24 hours of discovering the identity of a diseased cow in Washington state, the U.S. Department of Agriculture (USDA) had contacted over half the 578 businesses receiving parts of the cow. Consumers supposedly destroyed or returned 21,000 pounds of the 388,000 pounds of beef and bone (from one cow ?).

In early July 2004, press releases indicated that "federal secrecy rules" were interfering with the ability of Northwestern states of recall "tons of meat" that could have been from the cow in Washington state which had been found infected with "mad cow disease." Although federal officials knew which businesses had received the meat, they would not release this information because the states of Washington and

Oregon had not signed a pledge to withhold this information from the public. Federal law makes such recalls voluntary.

Eleven other states had signed the secrecy agreement.

The Consumers Union and others were asking Congress to pass laws to require the USDA to not only recall such hazardous meat products but also to share this information with the public. As on early July 2004 this had not occurred. Prion disease sources include deer and elk. Canada reported a second case of BSE in Ontario on January 3, 2005.

On January 26, 2004, new USDA "safeguards" were announced, but by July were not yet in place. These included ending the use of cow's blood fed to calves, the use of bovine brains or spinal cords in the manufacture of FDA regulated foods, dietary supplements or cosmetics. The fear was that the government could be overreacting. This is reminiscent of what occurred in Britain early in their "mad cow disease" (BSE) epidemic (40).

Some students of these prion diseases believe the potential exists for a more devastating global cause of death than AIDS.

At the risk of being too frightened, my readers could continue to follow the course of Bovine Spongiform Encephalopathy (BSE) and note the U.S. government's earlier disregard for precautions recommended by the World Health Organization.

EXHAUSTION OF NATURAL RESOURCES

[REFS 41-48]

Statistical studies in recent years had shown a drop in production of grain in proportion to population, a drop in supply of fresh water per person, depletion of ocean fisheries beyond sustainable limits, accelerating demands for scarce energy (especially in "developing" nations), human and industrial-wastes accumulation threatening human health. On the other hand, productivity and personal income continued to rise, especially in the "developing" world (41)(42)(43)(44)(45).

Over-fishing and pollution of the world's oceans, together with climate change, have created a crisis for fisheries.

There is near-extinction of some of the most valuable fish. In 1950, the world fish harvest was 21 million tons and in 1996 it was 116 million tons. Over-fishing is directly related to the human population explosion. One third of all fish species are threatened with extinction.

The Gulf of Mexico has a biological "dead area" the size of New Jersey due to "agricultural run-off" from the Mississippi River basin. Fertilizers spark huge blooms of algae which deplete the waters' oxygen. This has nearly killed bottom-dwelling marine organisms, causing fisherman to go further out to sea. Drinking water in several European countries is contaminated with fertilizer run-off.

Two out of every three species of birds are on a decline, with 11 percent "threatened" with extinction and 4 percent (403 species, "endangered" or "critically endangered." Another 14 percent are "vulnerable." Many have been poisoned by pesticides.

Of the almost 4,400 mammal species, about 11 percent are considered to be "endangered." Nearly half of our closest primate relatives, such as lemurs, monkeys and apes, are threatened with extinction, mostly due to hunting and habitat loss (46)(47).

Almost half of the forests once covering the earth are gone. Deforestation is expanding and accelerating. Between 1980 and 1995, at least 200 million hectares of forest vanished. This is an area the size of Mexico (48).

Deforestation and desertification are due to drought, logging, use of timber for firewood and clearing for farming.

This is especially true in Africa, Asia, Latin America, the Caribbean and the northern Mediterranean area. An estimated 2 acres of forest are lost per second. Climate change has made forests more susceptible to floods and drought. Some 12,000 years ago, the Sahara desert was green with lakes.

May 8, 2004, was United Nations *World Day to Combat Desertification and Drought.* A third of the planet's surface is threatened by the loss of fertile land and ever-rising temperatures. Climate change and pollution are leading to famine and poverty today. Legislation pending in the U.S. Senate will limit greenhouse gas emissions by U.S. polluters, to be effective by the year 2010. In 2003, nearly 20 Senators voted against this same legislation. Twenty were Republicans. The ten Democrats were from the Midwest and South.

Inquiries regarding the legislation may be directed to info@globalsolutions.org.

GLOBAL WARMING AND CLIMATE CHANGE

[REFS 49-52]

The ominous results of global warming, the accumulation of carbon dioxide and "greenhouse gasses," together with the increasing loss of the protecting "ozone layer" in the polar atmosphere, has been a continuing topic of concern in the United Nations General Assembly. This topic is discussed at length in the following reference (49).

As polar ice-caps melt, ocean levels rise. This has happened as each Ice Age subsided. Land bridges disappeared after being covered with water. Some of the more obvious include the Bering Strait, the English Channel and the Straits of Gibraltar. Studies of the ocean beaches are showing sand erosion of a few inches to 15 feet a year. In addition, beaches are increasingly closed due to the health hazards of fecal contamination. Over half of the U.S. population now live in coastal areas.

"Population growth is the biggest threat to our coastal areas," according to Walter McLeod, president of the Clean Beaches Council. (The new U.N. Secretary General, Ban Ki-Moon agreed – in 2007.)

In November 2004, in Iceland, the Arctic Council of eight nations agreed on a statement based on a report by 300 scientists, that a "more aggressive policy" was needed to counter global warming. Any specific mention of the U.S. contribution to the problem was avoided, allegedly for fear of "angering the Bush administration (50)(51)."

The Pentagon has reportedly warned that "there is substantial evidence that significant global warming will occur during the 21st century," leading to an abrupt climate change, which could destabilize the geopolitical environment, leading to skirmishes, battles, and even wars (52).

GLOBAL NUCLEAR HOLOCAUST

[REFS 53-57]

An early but most comprehensive evaluation of "lessons that need to be learned so the world can survive" may be found in the following reference, a 383 page compilation of articles published in the *Bulletin of the Atomic Scientists*. The Introduction to this book alone should horrify the most callous and greedy arms-maker and corrupt politician. An end to "civilization" *is* possible (53).

Reportedly, in the recent past, the U.S. was continuing to have over 10,000 nuclear weapons. Rather than defusing the threat of nuclear war, the presidential administration of George W. Bush was "pursuing the development of what are being called "*mini-nukes*," of 5 kilotons or less, more easily used in combat situations. It was believed that this will encourage other nations to follow suit. It seems significant that the U.S. Senate has not yet ratified the Comprehensive Test Ban Treaty (54).

Perhaps the most ominous prospect is for militant religious fundamentalists ("terrorists")to go beyond suicide-bombing, and to gain access to small *mini-nukes*. Other recent authoritative evaluations of the peril from misuse of nuclear power are in the following references (55)(56).

The United Nations General Assembly continues, year after year, to attempt to prevent the use of nuclear weapons as well as to address the critical issue of nuclear waste disposal (57).

NATURAL DISASTERS

[REFS 58-66]

In the last 500 million years there have been "many" reversals of the Earth's magnetic poles, corresponding to at least two major extinctions of life — at the end of the Permian Period, about 225 million years ago (MYA), and at the end of the Cretaceous Period, around 60-70 million years ago(MYA). Some galactic influence has been considered, as our solar system circumnavigates the galaxy (58).

The threat of mass extinctions – including most humans – lies in lack of preparation – based on historic events.

Let us now return to the effects of environmental pollution as a major factor in the current accelerating *extinction of species.* A *natural* or *background* rate of species extinctions over many millions of years before the appearance of humans has been estimated from the fossil records of marine invertebrates. This is on the order of one to three species a year. Currently the rate is believed to be "at least" 1,000 species a year.

Note here again that since the Industrial Revolution, the earth's environment has changed more than in the previous 4.5 billion years (59(60)(61)(62)(63) .

Despite the calculated "natural" or historical record of species extinctions before the appearance of humans, it should be noted that "large scale extinctions" had occurred, especially in the late Cretaceous Period—between 135 million years ago (MYA) and 70 MYA. The dinosaurs disappeared, along with a vast number of organisms, including all land vertebrates over 50 pounds. The causes continue to be debated today, but a meteor strike off the Yucatan coast is the prime suspect. Some of the other suspected causes include *global warming* and the *greenhouse effect*" and elevated amounts of ultraviolet radiation.

The largest of all mass extinctions of life were during the "Late Permian Period," some 250 million years ago, as the probably result of a "Mount Everest-sized meteor" striking the ocean northwest of Australia, creating a 125-mile wide crater. This was far worse than the mass extinctions of the "Late Cretaceous Period," some 65 MYA, which brought extinction to the dinosaurs after some 60 million years (64)(65).

At this point please note that an honored British astronomer, Sir Martin Rees, has reviewed the multiple sources for a possible "end to civilization" by the year 2100 AD/CE (66).

REFERENCES AND NOTATIONS: CHAPTER THREE

THE SEVEN MAJOR TIME-BOMBS

1) Reese,Martin. *Gloom and Doom by 2100*. New York: Scientific American, July 2004.

HUMAN POPULATION EXPLOSION

2) Tapinos,Georges, and Piotrow,Phyllis. *Six Billion People: Demographic Dilemmas and World Politics*. New York. McGraw-Hill, 1 978.
3) Lancaster, John. *Muslims protest population document*. LA Times – August 12, 1994.
4) Mesarovic,Mihailo; Pestel,Eduard. *Mankind at the Turning Point: Second Report to the Club of Rome*. New York. New American Library, 1974.
5) Freedman, Ronald ; Berelson, Bernard. *The Human Population*. San Francisco. W.H. Freeman and Company, 1974.
6) Brown,Lester. *The Twenty-Ninth Day: Accomodating Human Needs and Numbers to the Earth's Resources*. New York. W.W. Norton & Company, 1978.
7) Strickberger,Monroe. *Evolution – Third Edition*. Sudbury, Massachusetts, Jones and Bartlett Publishers, 2000. [See pages 554 and 635]
8) Cohen,Mark. *Carrying Capacity: How many are too many?* Amherst, NY. Free Inquiry, August-September 2004.
9) TIME Almanac-2000. *World Statistics*. [Pages 706-717]
10) Stark,Linda (Editor). *State of the World-2001*. New York. W.W. Norton, 2001. [See page 61]
11) Coale,Ansley. *History of the Human Population. Scientific American Book: The Human Population* W.H. Freeman, 1974.

CHEMICAL AND RADIATION POLLUTION

12) Krabill,H.F.; Mehlman, Myron (Editors).*Environmental Cancer – Volume Three*. New York and London. Halsted Press. John Wiley, 1977.
13) Barry,Brian. *Three Mile Island a million times over*. Portland, OR. *The Oregonian*, June 3, 2004.
14) Hebert, Josef. *Senate Oks relaxing cleanup for radioactive waste tanks*. The Oregonian, June 4, 2004.
15) King,Jonathan ;Rothman, Matt. *Troubled Water: The poisoning of America's drinking water – how government and industry allowed it to happen, and what you can do to ensure safe supply in the home*. Emmaus, Pennsylvania. Rondale Press, 1985.
16) Heilprin,John. Associated Press, November 26, 2004.

17) World Health Organization. *Neurotoxicity Risk Assessment for Human Health.* Geneva, 2001.
18) Environmental Health Criteria Series. World Health Organization, Geneva, 1976-1978. [Over 200 chemicals and environmental contaminants listed]
19) Kraybill,H.F.; Mehlman,Myron (Editors). *Environmental Cancer: Advances in Modern Technology – Volume Three.* [388 pages]
20) A Global Agenda: Issues Before the 56th General Assembly of the United Nations. 2001-2002 Edition. *New Convention on Persistent Organic Pollutants.*[See pages 130-131]
21) Choi,Charles. *Environment.* Scientific American, July 2004.
22) *Neurotoxicity Risk Assessment for Human Health: Principles and Approaches.* Environmental Health Criteria 223. Geneva. World Health Organization, 2001.
23) Ehrlich, Paul; Ehrlich, Anne; Holden, John. *Ecoscience: Population, Resources, Environment.* San Francisco, W.H. Freeman and Company, 1970, 1977. See pages 586-596.
24) Brunner, Borgna (Editor in Chief). TIME Almanac-2002. Boston, MA. See list of Human Carcinogens and Cancer Incidence rates, USA.[For an update, see TIME Almanac- 2006, pages 556-557]
25) Wikipedia at http://wikipedia.org/wiki/Agent_Orange
26) What Are Electromagnetic Fields? World Health Organization. Found at www.who.int/WhatisEMP/en/html]
27) ELF Electromagnetic Fields and the Risk of Cancer. See www.nrpb.org/publications/documents_of_nrpb/abstracts.
28) ELF Electromagnetic Fields and the Risk of Cancer. National Radiological Protection Board, Chilton, England. www.nrpb.org/publications/
29) Becker,Robert; Selden,Gary. *The Fields of Life.* New York. Ballantine, 1972.
30) Sulman,F.G. at Hebrew University, Jerusalem.

GLOBAL DISEASE PANDEMICS

31) *Mental Health-New* Understanding, *New Hope: The World Health Report-2001.* World Health Organization. Geneva, Switzerland, 2001.
32) TIME Almanac-2002.See page 132 for U.S. & page 557 for *World.*
33) World Resources Institute at www.wri.igc.org/wri using data from the 1996 study. Global Burden of Disease, Harvard University / World Health Organization.
34) Ayon-Shenker,Diana (Editor). *A Global Agenda: Issues Before the*

57th General Assembly of the United Nations. New York and Oxford. Rowman and Littlefield, 2002.[See pages 189-192]

35) Scientific American, July 2004.[See pages 8 and 86-93]
36) Prusner,Stanley (Editor). *Prion Biology and Diseases.* Cold Spring Harbor Press, 2004.
37) Yam,Philip. *Keeping the Mad Cows at Bay.* Scientific America, July 2002.]
38) Rampton,Sheldon; Stauber,John. *Mad Cow USA: Could the Nightmare Happen Here?* Monroe, Maine. Common Courage Press, 1977.
39) Mad Cow Home Page at www.mad-cow.org.
40) Dorkin, Andy. (07-01-04) and Cole, Michelle (07-02-04). Portland, OR. *The Oregonian* .

EXHAUSTION OF NATURAL RESOURCES

41) Starke,Linda (Editor). *Beyond Malthus: Sixteen Dimensions of the Population Pattern.* Washington, DC. Worldwide Institute, 1999.
42) King,Jonathan. *Troubled Water: The Poisoning of America's Drinking Water.* Emmaus, PA. Rondale Press, 1985.
43) Tapinos,Georges; Piotrow,Phyllis. *Six Billion People: Demographic Dilemmas and World Politics.* New York. McGraw-Hill, 1978.
44) Brown,Lester. *The Twenty-Ninth Day: Accommodating Human Needs and Numbers to the Earth's Resources.* New York. W.W. Norton & Company, 1978.
45) Kraybill,H.F; Mehlman,Myron (Editors). *Environmental Cancer: Advances in Modern Technology – Volume Three.*
46) Brown,Lester (Project Director); Starke,Linda (Editor). *State of the World-1998.* New York. Houghton Mifflin, 1964. [See Chapter 3,4 and 6]
47) Carson,Rachel. *Silent Spring.* New York. Houghton Mifflin, 1964.
48) Abramovitz,Janet. *Taking a Stand: Cultivating a New Relationship with the World's Forests.* Washington, DC. Worldwide Institute, 1998.

GLOBAL WARMING AND CLIMATE CHANGE

49) Ayton-Shenker, Diana (Editor).*A Global Agenda: Issues Before the 57th General Assembly of the United Nations.* New York and Oxford. Roman & Littlefield. [See pages 164- 176]
50) *Global Warming: Bulletins from a Warmer World.* Washington, DC. National Geographic magazine, September 2004. [NOTATION: A detailed presentation of the evidence for an accelerating global disaster – related clearly to human environmental pollution.

51) *Heating up at last ?* London, UK. The Economist, December 11-17, 2004.

52) *The Hightower Lowdown.* New York. May,2004. [lowdown@pipeline.com]

GLOBAL NUCLEAR HOLOCAUST

53) Ackland,Len;McGuire,Steven (Editors).*Assessing the Nuclear Age: Selections from the Bulletin of the Atomic Scientists.* Chicago,IL. Educational Foundation for Nuclear Science, 1986.

54) Mechtenberg,Molly. *Bush's New World Order.* Salem, OR. The Oregon Peace Worker. September, 2003.

55) Caldicott,Helen. *The New Nuclear Danger: George W. Bush's Military-Industrial Complex.* New York. The New Press, 2002.[Pages 48-49 *mini-nukes*]

56) Dweller,Nicole; Makhijani,Arjun;Burroughs,John(Editors). *Rule of Power or Rule of Law? An Assessment of U.S. Policies and Actions Regarding Security-related Treaties.* Produced by the Institute for Energy and Environmental Research and by the Lawyer's Committee on Nuclear Policy. New York. Apex Press, 2003.

57) Ayton-Shenker,Diana;Tessitore, John (Editors). *A Global Agenda: Issues Before the 56th General Assembly of the United Nation: An annual publication of the United Nations Association of the United States of America.* New York and Oxford. Rowman and Littlefield, 2002. [See part II: *Arms Control and Disarmament, Nuclear Matters*]

NATURAL DISASTERS

58) Magnetic Pole reversals. [See two refs below: www.science.nasa.gov/headlines/y2003/29dec_magneticfield.htm www.iceagenow.com/Magnetic_Reversal_Chart.htm.]

59) Starke, Linda (Editor). *State of the World-2001.* New York. W.W. Norton, 2001. [See pages 41-58]

60) Strickberger, Monroe. *Evolution – Third Edition.* Sudbury, Massachusetts, Jones and Bartlett Publishers, 2000. [See pages 429-431]

61) Landweber,Laura and Dodson,Andrew (Editors). *Genetics and the Extinction of the Species.* Princeton University Press,1999.

62) Carson,Rachel. *Silent Spring.* New York. H. Mifflin, 1964.

63) Brown,Lester. *The Twenty-ninth Day: Accommodating human needs and numbers to the earth's resources.* New York. W.W. Norton, 1978.

64) Choi,Charles. *Extinctions and Environment.* Scientific American, July 2004]

65) Strickberger, 2000. See pages 432-433. Also see en.wikipedia.org/wiki/Mass_extinction.

66) Rees,Martin. *Doom and Gloom by 2100.* Scientific American, July 2004. His book is titled *Our Final Hour,* 2003. In the United Kingdom the book is titled *Our Final Century.*

RIVAL RULERS OF THE WORLD

THE MAJOR CANDIDATES

[REFS 1-5]

The *Candidates* for *Rulers of the World* include the *Three Rival Religions* (Judaism, Christianity, Islam)as well as *Secret Societies* and *Shadow Governments* (Chapters 7 and 8).

The *Secret Societies* surely include the world's most influential organizations, such as the International Freemasons, the Skull and Bones fraternity at Yale University (an offshoot from the Bavarian Illuminati) and "the three most notorious secret societies". These have generally been thought to consist of the Trilateral Commission, the Council on Foreign Relations and the Bilderberg Group (1)(2)(3).

The behavior of many *candidates* for *Rulers of the World* is reminiscent of the *Art of War* principles attributed to the ancient, perhaps mythical, Chinese General, Sun-tsu (or perhaps the Russian General Antoine Henri Jomini). Consider, for example, the strategy of *spies everywhere* and pretending that your friends are your enemies and your enemies are your friends. *Blown cover for cover* has been a historical ruse to deceive the enemy. These "arts of war" continue today.

Deception and spying in both *hot* and *cold wars h*ave been the rule for centuries (4)(5).(Do grandiose *aims* ever justify any-and-all *means* ?)

Rival religions will be presented next under the history of each particular religion, including, Judaism, Catholicism, Islam, Christian-Protestantism and, perhaps, Communism. (Only the latter makes no claims of *divine* revelation.)

HISTORY OF THE RIVAL RELIGIONS

(REF 6)

This abbreviated history of religions will address only those of today's *Western World* – Judeo-Christian and Islamic. They are now the most involved in the ongoing global competition for the "minds of men." This section may almost require the reader to develop a *passion-for-history.*

Before the three current major modern religions appeared, the so-called *pagans* worshipped *nature,* but tended to attach human-like attributes to almost everything from the moon, sun, planets, stars and, even, the seasons).These were the gods (6).

The horrendous mutual massacres by persons with differing theories on the origins and meaning of life — in a timeless universe — were greatly intensified by the rise of Christianity and later, by Islam. These theories became powerful convictions and widely persist today. (*Pagans* still worship *nature.*)

EARLY JUDAISM

[REFS 7-19]

The origins of continuing belief in a single *God,* most exemplified by the Jewish, Christian and Islamic religions, usually leads to Abram of Ur – later called Abraham by a *God.*

Abram's birth in Sumer was in the first part of the reign of the "Divine" Sargon I (2350-2300 BC/BCE)- in the Early Bronze

Age. His birth is usually considered as anywhere between 3,100 and 2,000 BC/BCE. However, Sargon's regime fell to "barbarians" between 2,000 and 2,250 BC/BCE.

There seems to be major confusion regarding dates and Abraham. Compare dates in several writings (7)(8)(9).

It seems that, somewhere between the twentieth and nineteenth centuries BC/BCE, Abram of Ur (later called Abraham) took his family to Canaan by way of Haran (Turkey today), then settled in Hebron, now part of the West Bank in Israel. One source gives the date as around 1850 BC/BCE. His father, Terah, a pagan, had preceded him, migrating "to the west." In Palestine between 1300-1020 BC/BCE, monotheism was evident among the tribes of Israel, and again about 600 BC/BCE, in eastern Persia. In Palestine (Canaan) the historic Hebrew kings include Saul (c.1010 BC/BCE),David (c.1006-966 BC/BCE), and Solomon (966-926 BC/BCE).

The reasons are unclear for Abraham's departure at age 75 from Ur in Sumer, his birthplace, to travel to Canaan. There, in Shechem, Abraham had a "revelation," convincing him that he had made the right move. He then built an "altar for the Lord," considered a "holy place for generations (11)." The Old Testament traces Abraham's descendants to 560 BC/BCE.

Nebuchadnezzar II, emperor of Babylon, destroyed the Kingdom of Judea with the fall and destruction of Jerusalem (597-587 BC/BCE). Some former Palestinian Jews were held captive in Babylon, later scattered outside Palestine (the first diaspora) (12)(13).

The Assyrian Empire, existing from around 1800 BCE, fell in 609 BC/BCE, followed by the rise of Persia after 559 BC/BCE.

Historically, these kingdoms have been considered less important than "tiny" Athens, Sparta, and Judah.

Cyrus the Great, founder of the Persian Empire, conquered

Babylon 539 BC/BCE, then later assisted the Jews in returning to Jerusalem and Egypt.

Alexander the Great of Macedonia, conquered Jerusalem in 332 BC/BCE (14).

To provide a better general historical orientation: The Greek philosophers, included Plato(427-322 BC/BCE), followed by Aristotle, followed by Socrates (executed in 399 BC/BCE). The great Roman Senator, orator and writer, Marcus Tullius Cicero, was born in 106 BC/BCE, and was murdered in 43 BC/BCE.

After around 580 BC/BCE, when Cyrus the Great of Persia freed the Jews from captivity in Babylon, he allowed them to return to Jerusalem and Egypt. Empires and whole civilizations came and went. These included the Persians (559-330 BC/BCE), the Greek wars (500-400 BC/BCE), conquests by Alexander of Macedon (356-323 BC/BCE), and the Roman Empire (500 BC/BCE- 300 AD/CE). A very detailed chronology of these years is available elsewhere (15)(16).

Between 425-475 CE/AD, the Palestinian Talmud and, between 500-550 CE/AD , the Babylonian Talmud on Jewish law and tradition were written (17)(18).

For further information regarding early Judea, note that about 1947 AD/CE, ancient parchments were discovered in a cave high above the Dead Sea, some 13 miles east of Jerusalem. Of some 800 manuscripts, about a dozen were intact. They were reliably dated to the time of Jesus or soon afterward, when the Romans began persecuting Jews or in 70 AD/CE, when the Romans destroyed Jerusalem (19).

EARLY CHISTIANITY

[REFS 20-25]

An excellent chronology regarding the origins and course of Christianity from before the birth of Jesus around 4 BC/BCE and after his crucifixion around 30 AD/CE, is available (20)(21)(22).

There are other versions of how Christianity came about. They have been referred to as "Gnostic Scriptures" and "Lost Christianities (23)(24)."

One student of early Christians has estimated that in the year 40 AD/CE, there were only about 1,000 believers, but other estimates range in the many thousands by the year 100 AD/CE. By that date, there were a million Jews in Egypt, mostly in Alexandria. Christians were only one of many Jewish sects at the time (25).

Polytheistic *paganism* was "incredibly diverse and entrenched in the Roman Empire." Jews and Christians alike rejected the pagan religions of Rome. It was not uncommon for Jews and Christians to share a place of worship.

By around 150 AD/CE Christians were numerous enough to be considered a "Church" in the Roman Empire, with local bishops becoming the equivalent of regional imperial governors. They largely copied the political structure of the Roman Empire. They also now claimed succession from the apostle Peter.

In 165 AD/CE, a devastating disease epidemic, probably Bubonic Plague, was brought by troops returning from wars in the Mid-east (Parthia),lasting 15 years. It is believed to have facilitated Christian "conversions," because it raised doubts regarding the pagan gods. This was when it was noted that Christians had better survival rates.(They had no fear of death).

As "rulers of the known Western World," it seems of considerable importance to understand the transition from Roman Empire to *Holy Roman Empire* and from Emperors to Popes.

EARLY ROMAN CHURCH

[REFS 26-29]

In 306 AD/CE, Constantine, son of Emperor Augustus Constantinius, emperor of the Eastern Roman Empire, himself became emperor of the East and proceeded to "transform the Roman Empire, the Church and the position of the Jews."

In 312 AD/CE, Constantine, near Rome, with his troops exhausted and facing an entrenched rebellious army of the West, "saw a cross in the sky," above the legend "In Hoc Signo Vinces," meaning "in this sign you will conquer." His unexpected victory the next day, caused Constantine, his troops and, eventually, the empire, to become Christians and thereafter march under the banner of the Cross.

Constantine was baptized on his deathbed in 337 AD/CE. All his life he had been a pagan high priest of the sun god, Sol Inviticus.

This pagan religion was actually the "official" state religion until much later (26)(27).

Before Constantine, in 312 AD/CE, the early Christians comprised about one-tenth of the Empire's 50 to 60 million population and were only slowly growing.

Constantine, in 313 AD/CE, together with his brother-in-law and rival, Licinus, issued the Edict of Milan, which granted universal religious freedom to pagans, Christians and Jews. However, by 324 AD/CE, he "moved against pagans," as "slaves of superstition."

Although pagans were still in the majority, he proceeded to burn pagan temples and confiscated their treasuries, thereby "jettisoning the tolerance of Milan (28)."

Constantine, initially, did not want forced conversion for pagans and Jews, knowing that this was an impossibility, but instead wanted to win non-Christians over, in the same manner as he had (supposedly) been, in his vision of the cross. Never-the-less, he saw Christian diversity as a return to a kind of "polytheism." His goal of empire unification was "E pluribus unum" (one-out-of-many). He may also have believed the goal of unity, as a necessary *end*, could justify any *means*. To be religiously *different* became defined as *treason*, a political crime. (Recall Elizabeth I beheading her sister, Mary, Queen of Scots, over religious differences – called *treason*.)

In 325 AD/CE at the Council of Nicaea, Constantine achieved almost unanimous agreement from the 250 bishops assembled there. Essentially, it involved a statement of belief that "Jesus is God." This was approved by a vote. Those who dissented were exiled by Constantine. Still today, some Christians recite a somewhat altered Nicene Creed.

In 330 AD/CE, Byzantium was renamed Constantinople and was made "heathen Rome." (The final and permanent separation of the Western Church and the Eastern Orthodox Church at Constantinople wasn't until 1054 AD/CE).

Constantine was succeeded by his brother, Constantius II. Between 337-361 AD/CE Constantius made *Arianism* the binding religion of the empire. The Arianist Greek-speaking Christians, though strictly monotheistic, nevertheless, had been influenced by the Greek "atheist," Aristotle, who lived 384-322 BC/BCE. For this, they were detested by Catholics.

Ambrose (339-397 AD/CE), bishop of Milan, preached at the funeral of Emperor Theodosius in 395 AD/CE, that after all, why should those who dissented from defined dogma (Nicaea in 325) be allowed to live ? Ambrose might be considered the first of the Spanish Inquisitors, equating Jews to "Satan's surviving agents." (The first organized Inquisition, to eliminate *heretics* only began in 1231 AD/CE, when Pope Gregory set up roving Dominican and Franciscan ecclesiastic courts).

By 325 AD/CE, the Roman Empire was united under Constantine, with the Church leaders subservient to him (29).

After Constantine died in 337 AD/CE, "a bloody rivalry" ensued among his sons. In 361 AD/CE, Constantine's nephew, Julian, became emperor.

As the "last pagan emperor," Julian rejected Christianity, but was killed in Persia in 363 AD/CE. Emperors who followed him outlawed pagan worship, making "heresy" a capital crime.

Rome was invaded and "sacked" so many times, it is rather difficult to get all the dates straight. In 410 AD/CE, Visigoth hordes under Alaric sacked Rome, then invaded Spain in 412, causing the Vandals to flee to Africa. The last Roman (Western) emperor is considered to have been Romulus Augustus in 476 AD/CE. The last Catholic emperor of the East was Zeno, from 474-491 AD/CE. In 527, Justinian I, the Great, became the Byzantine emperor.

Constantine XI was the last Byzantine emperor when Constantinople was captured by Ottoman Turks in 1453 AD/CE. This ended the Christian Byzantine Empire after 1000 years.

HOLY ROMAN EMPIRE

[REFS 30-32]

It seems important to understand the sequence of events whereby the Roman Emperors, who initially barely tolerated Christians, gave way to Holy Roman Emperors, appointed and crowned by the popes.

It was said that, in the Roman Church, "the pursuit of power drives relentlessly among the unbroken shaft of apostolic succession," beginning with the apostle Peter, considered to be the first Pope (c.44 AD/CE) (30).

As background, we need to consider Charlemagne, the first Holy Roman Emperor. His grandfather, Charles Martel (c. 688-741 AD/CE), a Frankish leader, defeated the invading Muslims at Tours in 732 AD/CE, "saving the heart of Europe for Christianity." He also somehow "assisted" Pope Boniface in Germany (31) .

The grandson of Charles Martel, Charles I (c. 742-814 AD/CE), in 771 was crowned king of the Franks (of Germanic tribal origin), and was later known as Charlemagne. He united France, conquered Spain, Saxony and Bavaria by 788 AD/CE. Only Spain, Britain and Scandinavia eluded him.

On Christmas Day in 800 AD/CE, Charlemagne was crowned as the first Holy Roman Emperor by Pope Leo III(who reigned 795-816 AD/CE).This is believed to have guaranteed the pope's position in Europe, but lead to the *permanent* political division of the Catholic Church, between Western Rome and the Orthodox East in Constantinople.

After Otto I (the Great), Holy Roman Emperor (936-973 AD/CE), most of the emperors were German. Henry III, the Black (1039-1056 AD/CE), "strongest of the German emperors," made Germany a "feudal volcano," alienating bishops and nobles alike.

It was not just because emperors and popes both had felt they were "divine," as did the Pharaohs and the Augustus Caesars. It was pure greed. This seemed most evident in 1076 AD/CE at Canossa in the Apennine mountains. At that time, after having been excommunicated, the current Holy Roman Emperor, Henry IV, prostrated himself in the snow, arms outstretched in the snow in the sign of the cross, begging for absolution by Gregory VII, who was Pope from 1073-1085 AD/CE. For his humiliation, his excommunication was absolved.

His excommunication was the result of a "savage dispute" over who had the right to appoint abbots and bishops and thereby control vast lands and treasures.

This "humiliation" was believed to have prompted Henry's "brutal invasion" of Rome in 1083 AD/CE. However, in 1084, Pope Gregory's supposed Norman ally, Robert Guiscard, expelled Henry and, in the process, many atrocities were committed by his "motley army." Gregory fled to Salerno, where he died (32).

The balance of power between the emperor-kings and the Popes kept shifting for centuries, especially before the reign of

Innocent III (1198-1216 AD/CE), when the Church was at its most powerful point.

Frederick I (1123-1190 AD/CE), also known as Frederick Barbarossa, while the German Emperor, was the first to *claim* title to Holy Roman Emperor. He had sent his armies into Rome, in a sense isolating the Roman Church from the North. This probably helped create the "Reformation" in 1517 AD/CE in the heart of Europe. (See Martin Luther, 1483-1546.)

The period between 1254-1273 was considered to have been the end of the struggle between pope and emperor. Between around 1300-1400, German princes elected the Holy Roman emperor. Relations between Church and State had changed by the Reformation and Martin Luther (1483-1546 AD/CE) with his "95 theses."

Note that in Europe between 1347-1453 AD/CE at least 25 million people died from the "Black Death", bubonic plague. In 1453 AD/CE, Constantinople fell to the Ottoman Turks.

In 1438, Albert of Habsburg became emperor and the Habsburgs subsequently ruled the Holy Roman Empire for nearly 400 years.

Charles V (1500-1558), King of Spain, as part of the "Habsburg Empire" was elected Holy Roman Emperor in 1519. He was the last emperor to be crowned by a Pope, Clement VII, in the year 1530.In 1547,he scattered the Protestant League and a counter-revolution followed. The Thirty Years War(1618-1648)devastated Germany and left the Empire broken into hundreds of small principalities, mostly independent from the emperor (32).

In 1815, Austria emerged from the Congress of Vienna as Europe's dominant power, under the emperor, Franz Joseph I, who ruled from 1848 until he died in 1916.On November 11,1918, with the defeat of Austria, along with Germany, the Last Holy Roman Emperor, Charles, "stepped down but never abdicated.

THE CRUSADES

[REFS 33-43]

Nearly a dozen Christian "Crusades" occurred between 1096-1291 AD/CE (33). In 1095 AD/CE, Pope Urban II declared war against the "infidel" Muslims to repossess the Holy Land. Violence was defined as a sacred act. One motive was a hope to reunite the Western and Eastern churches.

The First Crusade (1096-1097) was the only one to be "successful". Crusaders did attack Jewish communities in the Rhineland. Many died rather than "convert". In 1099 "the main body of crusaders streamed back home".

The Christians captured Antioch and Jerusalem (1098-1099) and set up "Crusader Kingdoms", such as the Latin Kingdom of Jerusalem, making alliances with Muslim states and engaging intrigues among themselves. This "Latin government" depended upon later Crusades in 1148,1188,1202,1217,1228 and 1229 to prolong their occupation (34).

Despite a long history of antagonism of popes against the Jews, it should be noted that Pope Callixtus (1119) issued a papal bull in defense of the Jews and this was reissued by more than 20 popes during the next four centuries (35).

During the same period of the "First Crusade," five popular, aimless "mass migrations" emptied whole villages. One, under Peter the Hermit, involved some 7,000 and another, under Walter the Penniless, around 5000 AD/CE. Both reached Asia Minor and were annihilated.

Today, this must be considered as a monumental historical case of "mass –delusion" and/or "mass-hysteria." This was the same as the "Witch Craze" during the Inquisitions.

West and East relations had seemed improved in 1097 when Crusaders stopped by Constantinople and "swore the full feudal oath" to the Byzantine emperor Alexis 1. This favorable state of affairs persisted with the Greek emperor, Manuel 1, where his troops campaigned alongside the Latin armies in Egypt.

The Second Crusade began in 1144, led by the French king, Louis VIII. The breach between the western Latin and the eastern Greek Church had so widened that the Crusade leaders seriously considered the seizure of Constantinople. These Crusaders perished in Asia Minor in 1147.

In 1182, an uprising in Constantinople resulted in a "massacre of all the Latins" by a mob. Retaliation from the West came with the Fourth Crusade in 1203.

In 1184, Frederick Barbarossa, the first to *declare* himself Holy Roman Emperor, was preparing to lead the Third Crusade. At the Diet of Mainz, he met with 70 princes,70,000 knights and most of the German bishops. He wanted "the cathedral city on the Rhine" to become the second Rome. Actually, this consolidated the power of the Papacy as "a mass devotion to the cause of the Church", weakening the kings.

This Third Crusade (1189-1192,led by emperor Frederick Barbarossa, King Richard Lionheart of England and Philip II of France, was sparked by the capture of Jerusalem in 1187 by Saladin (1169-1193), the Egyptian Ayyubid Muslim ruler.

This Crusade ended with quarreling leaders and the drowning of the emperor Frederick Barbarossa. In 1199 Richard was killed in France.

Pope Innocent III(ruling 1198-1216), who made the *un-precedented claim* of absolute spiritual authority, "to judge all and be judged by no one", now possessed the universal authority as Christ's deputy, extending to an afterlife (36).

The Fourth Crusade (1200-1204)resulted in the "savage sacking" of Constantinopol by Latin knights in June 1203, who then established a Latin empire. Pope Innocent III wrote, "in profound indignation" (36):

"How can the Church of the Greeks be expected to return to devotion to the Apostolic See (Rome)when it has been seen the Latins setting an example of evil and the devil's work so that already, with good reason, the Greeks hate them worse than dogs."

The Albigensian-Waldensian Crusade (1208-1213) was proclaimed by Innocent III against these "heretics" in southern France. The victims were the Catharists of Albi and the followers of Peter Waldo, who had protested clerical corruption. Nobles supported the revolt, hoping to acquire Church lands.

Philip II (1180-1223), king of France, ignored the request for assistance from Innocent II.

Detailed descriptions of the Catharic heresy, the Catharists' devotion to John the Baptist and Mary Magdalene, and their close association the Knights Templar may be very enlightening. The Templars met death by torture for their supposed heresy between 1307-1314. Much has been written about the Templars and the early Freemasons.

The slaughter of over 100,000 Cathars in the Languedoc-Roussillon part of Providence, called the "first act of European genocide," was followed by the murder of every man, woman, child and priest in Beziers in July 1209. The number of dead was estimated at 15,000 to 20,000 (37)(38)(39)(40)(41).

The Children's Crusade (1212) resulted in only one of 30,000 French children and about 200 of 20,000 German children surviving to return home (42).

The Fifth (1217), Sixth (1228), Seventh (1248) and Eighth (1270) Crusades are described (43).

THE CATHOLIC INQUISITION

[REFS 44-49]

The Inquisition lasted between 1231 and 1834 AD/CE, ending

in Spain. In 1231, Pope Gregory IX issued the *Excommunicamus*, to begin 600 years of combating "heresy". This prolonged and agonizing story of human ignorance and greed and the willingness to torture and kill in the name of God – or in the name of divine Popes – has been documented in many places.

To attempt to summarize it here would add many pages.

It could be a worthy research project for a reader who values truth above comfort (44)(45)(46).

One aspect of the Inquisition was the "Witch-Craze" lasting 200-300 years and involving an estimated 40,000 to 100,000 trials and executions (usually by burning), eventually spreading to Puritan New England. In Salem, in 1692, they did not burn witches. They hung them (47).

The infamous guide to recognizing and extirpating witches was the *Malleus Malificarum (Witches Hammer)* published by the Dominicans in German in 1487, citing as authority the 1484 Bull issued by Innocent VIII (48)(49).

HISTORY OF ISLAM

[REFS 50-57]

Still another new religion joined the endless fray over *One-God* when Muhammad ibn Abdalla was born (c.570 AD/CE) in Mecca. His first divine revelation from the angel Gabriel was received in 610 AD/CE. He was eventually considered to be the last of the prophets.

Available in one unusual reference is a fascinating and intimate history of Muhammad's family, his revelations during seizures or trances, his many wives, and the beginning of Islamic civil wars following his death in 632 AD/CE (50).

Around 610 AD/CE, Mecca was a thriving Arabian center of trade. Rich and poor Arabs made their annual spiritual retreat to Mount Ramadan. However, old tribal values were being replaced by a new religion of "money." Each clan fought for a share of the new wealth. Among the Bedouin tribes, violence and vendettas as well as starvation were common. There was no central authority. They seemed doomed to constant warfare. Some Arabs in Syria had re-discovered the "authentic" religion of Abraham. They felt a great need for change.

In pre-Islamic Arabia, polygamy was common. Wives remained in their father's family. Most women had the same status as slaves. Female infanticide was also common. There were many pagan deities with their own shrines for worship. They had no notion of an after-life, believing only in fate and unquestioning obedience to the tribal chief or *sayyid.* Any "immortality" was not for the individual but only in the *survival of the tribe.* (At this point, it may be relevant to suggest that this attitude could be appropriate today, because immortality of our *species* should take precedent over any false promises of individual *immortality.*

Al-Lah,the High God of all gods was the same as the one worshipped by Muhammad's tribe, the Quraysh. The tribe had become very wealthy as traders and dedicated to "rampant and ruthless capitalism — the new religion of money."

As an Arab merchant in Mecca, Muhammad spent much time praying and had felt at one time that the survival of his own tribe, the Quraysh, was in jeopardy. In 510 AD/CE, while lacking sleep, an angel appeared and commanded him to "recite." Despite resisting, he finally found the first words of a new scripture pouring from his mouth. Thus began the *qur'an (Holy Koran)* (51).

Preaching in Mecca in 613, he was "reviled" and fled to Medina in 622 AD/CE. In 627, Meccans attacked Medina, but were repelled by Muhammad's 3,000 supporters. These new Muhamadans then accused a remaining Jewish tribe of "treason," executed the men and sold the women and children into slavery (52).

Around 622, Muhammad was devastated over his rejection by

the Jews of Medina. Thereafter, Muslims were told to face Mecca rather than Jerusalem when praying. This may be one early clue to a long lasting enmity towards Jews. At this time Muhammad developed the concept of jihad (holy war). This first leading to the conquest of Mecca in 630 AD/CE.

Muhammad's death in 632, was followed by civil war in 656 between his son-in-law, Ali, forerunner of the Shi'ite division of Islam, and the *puritanical* Kharijites (Seceders"). This latter Sunni group stressed *permanent religious aggression*. These Khawari Muslims conquered Syria and Iraq (633-637). Iran (Persia) and Egypt in 642, Armenia in 653 and Afghanistan in 644, "ruling as an autocratic but tolerant minority." (The Shi'ite leader, Ali was murdered in 661.)Muslims invaded Spain and the Indus Valley in 720-711. Today, over 80% of Muslims are the militant Sunni (53).

The Islamic "Civil wars" following Muhammad's death and the controversy between Shi'ites and Sunnis over who should be his legitimate successor continued for centuries. The first caliph, Abu Bakr (632-234 AD/CE), was the prophet's close friend. The Third caliph, Umar ibn Affan (644-656 AD/CE), a member of the Amayyad clan (Sunnis), was murdered by "rebellious Egyptian forces."

The fourth caliph was Ali ibn Abu Talib in 656 AD/CE. He was the first cousin and son-in-law (married Fatima) of the Prophet and reigned until he was murdered in 661 AD/CE by a member of the Umayyad clan (later known as Sunnis). Thus began the Ummayyad dynasty, which lasted until 750 AD/CE The Umayyads were the first of the orthodox *Sunni* majority.

Later, the followers of Ali were known as *Shi'ites*, or Shi'ate Ali (partisans of Ali), a minority Muslim faction still today. Some 88 percent of the world's 1.3 billion Muslims are Sunnis. Iran, with 90 % Shi'ites, controlled by powerful clerics,worries the Sunnis of Jordan, Turkey and Saudi Arabia.

Shi'ite opposition against the Umayyas began when Ali, in 661, and then his son, Husayn, in 687 AD/CE, were murdered. Husayn's head was sent to Yazid, the Umayyad caliph in Damascus.) A shrine to Ali in Najaf, Iraq, is one of the holiest sites for Shi'ites. He and his son, Husayn, are buried in Karbala, Iraq. A Shi'ite believes that being buried nearby increases the chance of Paradise (54).

Between 744–750 AD/CE, another Islamic civil war ended with the defeat of the Umayyads by the Abbasids. Al-Abbas , the Prophet's uncle, rebelled against the "haughty, powerful, impious, worldly and extravagant" Umayyads and was proclaimed caliph. Most of the Umayyads were "liquidated," by an Abbasid coalition which included Shi'ites. The Abbasid Empire began when al-Abbas proclaimed himself caliph, thus began the Arab Umayyad (Orthodox Sunni) dynasty (55).

The first Arab attack on Constantinople, capital of the eastern Byzantine Roman empire, had been in 669 AD/CE. In 1453 Muslim Turks conquered that city, ending the Greek orthodox Byzantine Empire and beginning the Muslim Ottoman Empire.

A chronology of the Arab (Islamic) Empire, with almost endless wars involving Muslims, Christians, Jews and Mongols is described in detail elsewhere (56)(57).

PROTESTANT REFORMATION

[REFS 58-65]

The unsuccessful Catholic Reformation could have started when the 34 year old Augustinian monk, Martin Luther (1483-1546 AD/CE) posted his "Ninety-five Theses" (against the Catholic "indulgences")on the door of the castle church in Wittenberg, Germany in 1517.

At this time, the Roman Catholic Church was still obsessed with eliminating "heresy" and was weakened by many setbacks. A few of these included the "humiliating" seizure of Pope Boniface VIII in 1303 by the French king, Philip IV, and the translation of the Bible from Latin into English by John Wycliff between 1376-1382; the "Great Schism" between rival popes in Rome and Avignon, France, lasting from 1378 to 1417; and then the first printed Bible by Johann Gutenberg in 1455 (58)(59).

Martin Luther had an almost life-long struggle with an overwhelming fear of death. Doubts regarding his own "life-after-death" began when, as a youth, he was terrified during a violent storm. He vowed then to become a monk.

In 1520 he had written "Theology of the Cross" and in 1523 wrote "Jesus was a Jew." He was a "stout defender" of Jews. His hero was Saint Augustine (354-430 AD/CE) (60).

His dread of death worsened around 1543 when his beloved daughter, Magdalena, died in his arms. Shortly before this he had written "On Jews and Their Lies".

Rejected by Catholics and Jews, he spent his life as an angry man "a rabid anti-Semite, convulsed with a loathing and horror of sexuality and believing that all rebellious peasants should be killed" (61).

Luther's disenchantment with Jews became a "venomous hatred". He regarded both Catholics and Jews as his mortal enemies. Catholics blamed Jews for the spread of "heresy", because they rejected Jesus (as .... a son of God, born of a virgin and resurrected) and refused to convert. He was quoted as saying that Jews contradicted the one thing that kept him from going mad – meaning the promise of a life-after-death (62).

In 1519, the last Holy Roman Emperor to be crowned by a pope was Charles V. He had defended Jews, but exhausted by a succession of wars and the futility of trying to turn back Protestantism, abdicated in 1556 and died in a monastery two years later. Popes continued to blame Jews for the Christian "heretics".

On Martin Luther's 455th birthday (November 10,1938), synagogues were burning in Germany – *Kristallnacht.* His wish for a German *judenrein* was only beginning.

Luther also was "a firm believer in witchcraft." A Papal Bull in 1484 by Innocent VIII began the "Witch-Craze" in Europe, with "hideous persecution, thousands of men and women tortured until they confessed to astonishing crimes" ((63)(64).

In 1487,the *Malleus Maleficarum (Witches Hammer)* was published by a Dominican inquisitor, Heinrich Institoris, with the full backing of Rome. In New England, some Puritans preached that they were under attack from witches as a punishment from God. The horrors of Salem, Massachusetts, in 1692 were such as to put an end to witch hunting.

There were "witches" (as well as gods and goddesses) and magic, in ancient Greece, in Rome for hundreds of years before Christianity and among the Germanic, Norse and Celtic peoples (65).

RELIGIONS BY THE NUMBERS

[REFS 66-73]

The major world religions include the Roman Catholic (Universal) Church, Orthodox (Eastern) Catholics, Protestant and other Christians, Muslims (Islamic World or Khilafah), Jews, Sikhs, Hindus, Buddhists and Shintoists. Perhaps Atheists and Communists should be included as religions, although they allegedly adhere to the methods of science for truth rather than the revelations described in Holy Scriptures.

Some religions still consider persons outside their own belief system to be *infidels or heretics* and continue to attempt to *covert* them. In past centuries the non-believers often have had the alternative of being tortured, killed or *expelled*. Today, this is exemplified by the Islamic Mujahidin.

In the United States, some 23 major U.S. denominations with membership under 5 million but over 500,000 are listed (66).

The percentages of each religion by country is also available in the reference given below. For the most part, the countries which are overwhelmingly a single religion are Catholic or Muslim. It may be noted that China's "official" religion is "atheist," but many Chinese may be considered as followers of Confucianism, Taoism or Buddhism (67).

The estimated number of *Atheists* world-wide has been given as 149,913,000 (2.5%) mostly in Asia. My readers are encouraged to research *atheism* and *humanism* (68-73).

Another valuable statistical source regarding religions, socio-economics and politics in the Middle-East might be examined and compared with the above information. Much additional information is needed here regarding the *animistic* religions prevailing today, especially in Africa. Some continue to be involved in horrendous *crimes-against-humanity* (74).

REFERENCES AND NOTATIONS: CHAPTER FOUR

THE MAJOR CANDIDATES

1) Marrs,Jim. *Rule by Secrecy.* New Part One: Modern Secret Societies] York.HarperCollins, 2000.
2) Ross,Robert Gaylon,Sr. *Who's Who of the Elite: Members of the Bilderbergs,, Council of Foreign relations & Trilateral Commission.* Spicewood TX. RIE Publishers, 2000.
3) McManus,John. *The Insiders.* Appleton,WI. The John Birch Society,1992.
4) Owen,David. *Hidden Secrets: A Complete History of Espionage.*Toronto, Canada. Firefly Books,2002.
5) Perisco,Joseph. *Roosevelt's Secret War: FDR and WWI Espionage.* New York. Random House, 2001.

HISTORY OF THE RIVAL RELIGIONS

6) Garraty & Gay.*Columbia History of the World,*1972.New York. Harper & Row, 1972.[See Chapter 8: Of Gods and Men]

EARLY JUDAISM

7) Garraty & Gay. *Columbia History of the World,* 1972.[See Pages 161-163 and Chapters 25 and 35]
8) Shanks, Hershel. *The Mystery and Meaning of the Dead Sea Scrolls.* New York. Random House,1998
9) Lewis, Brenda (General Editor). *Great Civilizations.* Bath, UK. Parragon Publishing, 1999. [See page 148]
10) www.greatpyramid.net
11) Armstrong, Karen. *A History of God,*1993.
12) Garraty and Gay. *Columbia History of the World,*1972. [Seepages 162 and 418]
13) Stearns,Peter (General Editor). *The Encyclopedia of World History. Sixth Edition.* New York. Houghton Mifflin Company,2001.
14) Garraty and Gay. *Columbia History of the World,* 1972. New York, London. Harper & Row,1972. [See pages 161-163 and Chapters 25 and 35]
15) Brunner, Borgna (Editor in Chief). TIME Almanac 2002. Boston, MA, 2001. ["Headline History," pages 671-70]
16) Kindar,Hermann; Hilgemann,Werner. *The Anchor Atlas of World History Volume One. (*Translated). Garden City, NY. Anchor Books/ Doubleday,1964.[Pages 37-39]
17) Stearns Peter (General Editor). *The Encyclopedia of World History. Sixth Edition.* New York. Houghton Mifflin Company,2001.

18) Garraty & Gay (Editors).*Columbia History of the World.* [See Chapters 161-163 and Chapters 25 and 35]
19) Shanks,Hershel. *The Mystery and Meaning of the Dead Sea Scrolls.* New York. Random House,1998.

EARLY CHRISTIANITY

20) Carroll,James. *Constantine's Sword,*2001.[See Chronology]
21) Shanks,Hershel;Witherington III,Ben. The Brother of Jesus. HarperSanFrancisco,2003.
22) Reuchlin, Albelard. *The True Identity of the New Testament.* Kent, WA. Abelard Reuchlin Foundation, 1979.
23) Layton,Bentley. *The Gnostic Scriptures.* New York. Doubleday,1987.
24) Ehrman,Bart D. *Lost Christianities.* Oxford University Press,2003.
25) Stark,Rodney. *The Rise of Christianity: How the Obscure, Marginal Jesus Movement became the Dominant Religious Force on the Western World in a Few Generations.* Harper SanFrancisco, 1997.Copyright by Princeton University Press, 1996.

EARLY ROMAN CHURCH

26) Carroll,James. *Constantine's Sword: The Church and the Jews* .New York .Houghton Mifflin, 2001. [NOTE:See Chapter 19 and pages 180-192 re Sol Inviticus, Constantine's pagan religion – as it's high priest.]
27) Baigent,Michael;Leigh,Richard;Lincoln,Henry. *Holy Blood, Holy Grail.* New York. Dell-Bantam Doubleday, 1982,1983. [NOTATION: See pages 365-368]
28) Carroll,James. *Constantine's Sword: The Church and the Jews.* New York. Houghton Mifflin, 2001.[NOTATION: See Chapter 17: The Story of Constantine]
29) Carroll,James. *Constantine's Sword: The Church and the Jews.* New York. Houghton Mifflin, 2001.[NOTATION: See pages 280-282.]

HOLY ROMAN EMPIRE

30) Carroll,James. *Constantine's Sword: The Church and the Jews.*New York. Houghton Mifflin, 2001.[NOTATION: See Chapter 57: *The Church and Power.* Also see pages 238-240]
31) Stearns,Peter (General Editor).*The Encyclopedia of World History – Sixth Edition* .New York. Houghton Mifflin, 2001. [NOTATION: See pages 205,213,664.]
32) Brunner, Borga. (Editor in Chief).TIME Almanac- 2002. [NOTATION: See *Germany,* pages 773-775]

THE CRUSADES

33) TIME Almanac- 2006. *Headline History*. Boston, MA. [NOTATION: See pages 667-668]

34) Stearns,Peter (General Editor).*The Encyclopedia of World History –Sixth Edition* .New York. Houghton Mifflin, 2001. [NOTATION: See pages 128,233]

35) Carroll,James. *Constantine's Sword: The Church and the Jews.* New York. Houghton Mifflin, 2001.[NOTATION: See *Chronology*, page 624]

36) Carroll,James. *Constantine's Sword: The Church and the Jews.* New York. Houghton Mifflin, 2001.[NOTATION: See page 281 regarding Pope Innocent III on an *after-life*]

37) Garraty & Gay (Editors).*Columbia History of the World*,1972. [NOTATION: See page 453 for Pope Innocent III and his remarks about the Fourth Crusade and the sacking and looting of the Greek Orthodox Constantinopol.]

38) MacKenzie,Norman (Editor).*Secret Societies.* New York. Crescent Books/Crown Publishers, 1967.[NOTATION: See Chapter Five: *The Assassins and the Knights Templar*]

39) Garraty & Gay (Editors).*Columbia History of the World*,1972. [See pages 389, 421 (Cathars)]

40) Picknett,Lynn;Prince,Clive. *The Templar Revelation.* New York. Simon and Schuster,1997.[*Chapter Four: Heartland of Heresy.*]

41) Wassemann,James. *The Templars and the Assassins.* Rochester, VT. Inner Traditions,2001.[See Chapter 19: The Cathars and the Albigensian Crusade]

42) Ridley,Jasper. *The Freemasons: A History of the World's Most Powerful Secret Society.* New York.Arcade,1999,2001. [See *Chapter 2: The Heretics.*]

43) TIME Almanac-2006. See footnote page 667: *The Crusades* (1096-1291).

THE CATHOLIC INQUISITION

44) TIME Almanac-2006. [The Inquisition (1231-1834) ending in Spain on pp 668,670]

45) Carroll,James.*Constantine's Sword*,2001.[See *Part Five: The Inquisition*]

46) Langer,William. *The Encyclopedia of World History – 6th Edition.*

47) Streeter,Michael. *Witchcraft: A Secret History.* London. Quarto/Barrons,2002.

48) Armstrong,Karen.*History of God*, 1993.[See page 323]

49) Stearn,Peter.*The Encyclopedia of World History*, 2001.

HISTORY OF ISLAM

50) Caner,Ergun;Caner,Ermir. *Unveiling Islam: An Insider's Look at Muslin Life and Beliefs.* Grand Rapids, MI. Kregel Publications, 2002.

51) *Armstrong,Karen. A History of God: The 4,000-Year Quest of Judaism, Christianity and Islam.* New York. Ballantine Books,1993. [Chapter 5: Unity – the God of Islam]

52) Stearns,Peter. *The Encyclopedia of World History.* New York. Houghton Mifflin. 2001.

53) TIME almanac-2006.[See C*ountries of the World, pages 718- 906,* for Islam in each country]

54) Di Giovanni,Janine. *Shiites of Iraq: Reaching for Power.* Washington, DC National Geographic magazine, June 2004, pages 2-35. [An excellent comprehensive update]

55)Garraty and Gay. *Columbia History of the World,* 1972.[See pages 253-289 (Arabs and Islam) and pages 272-273]

56) Stearns,Peter, *The Encyclopedia of World History.* New York. Houghton Mifflin, 2001.[*The Rise and Expansion of Islam, 610-945; Middle East and North Africa, 945-1500; South and Southeast Asia, 500-1199 and 900-1557*]

57) Armstrong,Karen. *A History of God: The 4,000-Year Quest of Judaism, Christianity and Islam.* New York. Ballantine Books,1993.[See pages 258 (Inquisition) and 196-204 (Crusades)]

PROTESTANT REFORMATION

58) Garraty and Gay (Editors). Columbia History of the World, 1972.[Chapters 42 & 43: *The Reformation ;* Chapter 44: Counter-revolution]

59) Stearns, Peter. The Encyclopedia of World History – Sixth Edition, 2001. [See "Reformation," pages 283,285,300,309]

60) Carroll, James. Constantine's Sword, 2001.[See Chapter 21, *Augustine Trembling* and pages 366-368 re Luther's anti- Semitic influence for centuries in Christian Germany]

61) Armstrong, Karen. A History of God, 1993.[Pages 275-280 re Luther and Calvin]

62) Carroll,James. Constantines Sword 2001.[See pages 366- 370]

63) Armstrong, Karen. *A History of God,* 1993.[Page 275]

64) Stearns,Peter. *The Encyclopedia of World History,* 2001. [Pages 284,323,403,411]

65) Streeter,Michael.*Witchcraft: A Secret History.* London and New York. Barrons,2002.[See Chapter 8: *Salem-A Town Possessed.*

RELIGIONS BY THE NUMBERS

66) TIME Almanac 2006 .[See pages 360-367: World Religions and Largest U.S. Churches in 2004. NOTE: In the 2002 TIME Almanac the estimated world total *Atheists* were 149,913,000 or 2.5 % - mostly in Asia. The word *atheist* does not appear in the 2006 TIME Almanac]

67) TIME Almanac 2006. [See *Countries of the World*, pages718- 798, which includes percentage of Religions]

68) Joshi. S.T. (Editor). *Atheism.* Amherst, NY. Prometheus Books,2000.

69) Armstrong, Karen. *A History of God*, 1993. [See *Introduction*, pages xix and xx]

70) Cherry, Matt; Flynn, Tom; Madigan, Timothy (Editors), *Imagine There's No Heaven.* Amherst, NY. Council for Secular Humanism. 1997.

71) TIME Almanac-2006. [*Countries of the World*, pages 718- 798]

72) Mulroy,Kevin (Editor-in-Chief). *Atlas of the MiddleEast.*Washington, DC. National Geographic Society, 2003. [An unusually beautiful book with colorful maps]

73) Igwe,Leo. *Ritual Killing and Pseudoscience in Nigeria.* Amherst, NY. Skeptical Briefs, June 2000. 104

CHAPTER FIVE

HOLY WARS AND HOLY WARRIORS

THE HOLY WARRIORS

[REFS 1-4]

Major religions competing to be *Rulers of the World* surely include the Khilafah (Islamic World) and the Roman Catholic (Universal) Church. If *religious zeal* is the hall-mark, perhaps "Communism" and "New World Order" advocates should be included, despite the absence of *divine revelation.*

It seems clear that militant Islamists (Mujahidin) and the early Catholic warriors (Jesuits) have been long-term enemies of Judaism.

Although Jerusalem had fallen to the militant Muslims by 637 AD/CE, it seems unlikely that Muslim enmity was never so horrendous as that of Catholicism during the Crusades and through the Inquisitions.

Consider the atrocities against Rhineland Jews during the First "Christian" Crusade (1096 AD/CE) at the call of Pope Urban II (1095 AD/CE).

This included the goal of purging the "Christ-killers." However, Urban II said this was "not anticipated."

These Crusades made the Holy See in Rome, in effect, the ruler of Europe (1)(2)(3).

Especially within the past century, this anti-Semitism was with tacit approval of the Holy See and Pius XII. For centuries the Western Roman Church had sought realignment with the Eastern Orthodox Church. This could have been related to Hitler's mindless Nazi invasion of Russia (4).

CATHOLICS VERSUS JEWS

[REFS 5-8]

Again, examples of the Catholic position when Pope Pius XII, in his 1933 AD/CE Concordant, approved of Hitler's goals. Later, knowing of the death camps, he stated that National Socialism was less a threat than communism. (Does the current Pope also credit "communism" to the Jews?) On the other hand, Hitler, in his "ravings" near the collapse of Germany, spoke of his plan, if he lost the war, to "hang the Pope."(5)(6)(7)(8).

THE SOCIETY OF JESUS (THE JESUITS)

[REFS 9-20]

The convoluted, turbulent and violent history of this "secret army of the Papacy", has included repeated efforts at times to disband the Order and their expulsion many times from many countries world-wide. This has been explained by their fanatical efforts to convert or to destroy the opponents of the Roman Church. Revolutions and massacres have been attributed to their unrelenting efforts. This followed a pattern of "the end justifies (any) means" (9)(10)(11)(12).

The repeated shifts of allegiance to the elected Pope over the centuries seemingly resist understanding. Despite oaths of total obedience to the Holy See, from its founding by Ignatius of Loyola in 1534 AD/CE, at least two Popes, Clement XIII and Clement XIV, were assassinated (poisoned), allegedly by Jesuits — in 1769 and 1774 AD/CE. These two Popes had finally succumbed to incessant demands to "suppress" the Jesuits. As a result, Lorenzo Ricci, the 18th Superior General for the Society of Jesus, was imprisoned and died a few years later. The Jesuits "appeared to submit" to the Pope's Brief of Dissolution but they wrote "innumerable pamphlets *against* the pope and incited rebellion."(13)(14)(15).

Despite the abolishment of the Order in 1773 AD/CE by Benedict XIV, the Superiors General of the Society of Jesus are still listed from Ignatius Loyola (1541 AD/CE) to Peter Hans Kolvenbach in Russia (1993 AD/CE). This listing includes some 33 Superiors General of the Society of Jesus – described as the "Black Popes" (16)(17).

Also, a listing is available for 240 popes, from St. Peter (around 42-67 AD/CE) to Benedict XVI (2005)to present) (18).

Jesuits were global missionaries by 1556 AD/CE, the year of Loyola's death, trying to convert the "pagans" of India, China, Japan, the New World and especially, France, Spain, Portugal, Italy, and England (by way of Ireland). Incidental to their saving souls, they greatly enriched the Vatican's coffers. Over the centuries they were, sooner or later, persecuted, hung or expelled from many nations as extreme troublemakers (19).

In 1873 AD/CE, in France, an attempt was made to restore a king who might be favorable to the Church, but this attempt failed. Priests "slandered the Republic" as a government of free-thinkers and Masons. However, many Catholics found the priest to be a "troublesome man," interfering with their independence.

In 1886, General Georges Boulanger was French Minister for War and "was looking like a future dictator." He made a secret agreement with Royalist Catholic members of parliament, but he was defeated in the polls in 1889. Boulanger then sent a letter to Pope Leo XIII saying that when he (Boulanger) "held the sword of France in his hands" he would assure the rights of the Church.

In December 1894, Alfred Dreyfus, a Jewish captain in the artillery, was speedily convicted of treason on a trumped up charge of giving defense documents to the German embassy. This resulted in an "anti-Semitic fever." On Devil's Island Dreyfus was tortured. Eventually he was exonerated on all charges (20).

The ambitions of the Roman Church have been described at length in several publications, often in relation to Freemasonry, the Jesuits and the New World Order.

The activities of the Jesuits for over four and a half centuries can not even begin to be told in these few pages.

The reader is invited to do their own research, e.g., the Jesuit-Jacobin-Freemasonic "alliance."

ISLAMIC MUJAHIDEEN

[REFS 21-37]

Islamic warriors are known as *mujahideen*. Also note that *jihad* refers to *struggle* or *holy war* and *khilafah* means *Islamic World* (21)(22)(23) ."

A recent editorial provides a succinct example of the fanatical ideology of Saudi Arabian Muslims—kill all "infidels" and "Westerners." This is the world view taught to children in Arab schools today—financed by Saudi princes with old money. The Saudi princes are believed to be heavily obligated to the radical Wahhabi clerics, who have waged *jihad* since as early as 1811 AD/CE (24)(25).

In regard to *Islamic warriors*, one should consider "The Charter of Allah: The Platform of the Islamic Resistance Movement (Hamas)." In the Articles of this declaration, it is made clear that "the Land of Palestine had been an Islamic *Waqf* throughout the generations" and that various organizations "which take all sort of names and shapes such as the Freemasons, Rotary Clubs ...gangs of spies ..nests of saboteurs . . and are "implementing Zionist goals (25)."

The term "Waqf" refers to "an old tradition" of Islam whereby some asset is "dedicated to a pious purpose." The donor may be a ruler, a government or a private individual. In the Hamas "Platform," "...the land of Palestine belongs to Islam until the "Day of Resurrection," as is true of "all lands conquered by Islam by force." Article Twelve states "Hamas regards Nationalism (Wataniyya) as part and parcel of the religious faith. Nothing is loftier or deeper in Nationalism than waging Jihad against an enemy and confronting him when he sets foot on the land of the Muslims." It seems noteworthy that Palestine was promised by God to the Jews and by Allah to the Muslims (26).

The global militant Islamic network, Al-Qaeda (Arabic for "the base") was apparently originated by Osama bin Laden during the Soviet occupation of Afghanistan. The main objective of Al-Qaeda has been stated as "the destruction of all infidel nations that appear to oppose their goal of a pure Islamic state" (27).

The Koran and history leave little doubt regarding the requirement to convert or destroy the non-believing "infidel." The Arab Empire was a conquest by the sword, beginning with Muhammad's victorious return to Mecca, "wiping out all opposition," two years before his death (632 AD/CE).

In modern times, the "World Islamic Front for Jihad against Jews and Crusaders" was organized in 1998 AD/CE by a Sunni Islamic faction. This was with a commitment to "kill the Americans, civilian and military, in retaliation for any U.S. attack on Iraq or any demonstrated hostility anywhere else in the Muslim world." One recent author has presented a carefully researched and analyzed assessment of the militant Muslim attack on the United States Pentagon and the World Trade Center on September 11, 2001. To some persons' surprise, it was not about American pornography or alcohol, but that his message was against U.S. foreign policies. These centered on the continued presence of American military in "holy" Saudi Arabia and failure to provide reasonable restraints on Israeli Ariel Sharon's treatment of Palestinians in "occupied territory (28)."

Perhaps only a minority of Muslims are religious extremists, but "Almost every Islamic country has its militant faction, often two or three" and ".. all share the same goal of an Islamic world, or, as they refer to it, a *Khilafah*. In such a world there would be no separation of church (religion) and state. "A concurrent intensification and diversification of the Islamic buildup has taken place at the heart of the West, Western Europe, the United States and Canada — not only for the support of spectacular terrorist strikes but also to accelerate the ongoing erosion of Western Society." ".. a new generation of networks is already operating throughout the West completely outside the realm usually associated with Islamic terrorism"(29)(30)(31).

Bin Laden surely was trying to make a "statement" in choosing the New York twin trade towers of the World Trade Organization as his target on September 11, 2001. He has been quoted at one time as saying that Muslims were tired of being treated like insects, possibly referring to the Soviet invasion of Afghanistan and the expansion of Israel into Palestinian territory or, perhaps, corporate exploitation of Muslim countries (32)(33)(34)(35).

In 2005 AD/CE, after the Anglo-American invasion of Iraq, an immense problem for the United Nations was how to assist that country's restoration to its status as a functioning and newly democratic member of the United Nations. Related current issues for the United Nations agenda must continue to include the Taliban, still in Afghanistan, and global Islamic terrorism.

The United Nations agenda 2001-2002, under the heading of "Keeping the Peace" in regional conflicts, included an in-depth analysis of the Israeli-Palestinian conflict, the Israeli "occupied territories," as well as Islamic terrorism in Afghanistan and Indonesia. This topic appears in each annual UN Agenda (36).

In late 2004, in Darfur, wholesale genocide was continued by the Arab *Janjaweed* militia backed by the Sudanese government.

This followed earlier protests by black farmers over unfair government land and water policies. So far, the Western reaction had been lukewarm at best (37).

MUSLIMS VERSUS JEWS

[REFS 38-42]

In the Mid-East, the continuing *time-bomb* of Muslims versus Jews justifies a somewhat detailed review of its history. This has a bearing on the ancient dream of many Jews of returning to their homeland in Palestine was well as the ancient alienation among the "children of Abraham," Isaac and Ishmael. This subject has been written about for hundreds of years. The dates often disagree and mythology is rampant in the earliest writings regarding creation of the world, mankind and the gods. Abraham could have made his journey from Ur to Chaldea to Canaan somewhere between 2000-1900 BC/BCE (38).

Both Jewish and Islamic religions "glorify" Abraham. The great-divide of these two major religions has been partly attributed to the controversy over which of his two sons, Isaac or Ishmael, was to be Abraham's rightful heir. This included the *myth* of God's promise of Abraham fathering "many nations as numerous as the stars," and that, as God's "chosen people," they would one day possess all of Canaan.

Isaac (later renamed Israel), was born to Abraham and Sarah, who was past menopause and "barren." As the first-born, he was considered to be the offspring of a god and was to be sacrificed. Although this occurred on Mount Moriah in Jerusalem, it was much like a pagan custom in Babylon.

Legend has it that God relented at the last minute and Isaac was spared to receive Abraham's promised inheritance.

To comfort Abraham, God promised Ishmael, son of Abraham's concubine, Hagar, that he, also, would be the father of a great nation. Abraham is said to have left Hagar and Ishmael in the valley of Mecca. Later, Abraham and Ishmael built the temple, Ka'abah. Ishmael then became the father of the Arabs (39).

It was only after some thousand years, around 622 AD/CE, Mohammed ibn Abdullah had learned from friendly Jews in Medina that another son, Ishmael, was born to Abraham's concubine, Hagar. Sarah had demanded that Abraham "get rid of" Hagar and Ishmael. Instead, Abraham had, according to Islamic tradition, brought them to Mecca (40)(41)(42).

Another early source of the alienation of Islam from Judaism was in Medina, after Mohammad's "migration" (hijra) from Mecca around 622 AD/CE. In Medina the Jews initially listened with interest but later laughed and scoffed at his new religion, saying there were no more prophets, only the coming of a Messiah. They joined the pagans who were also hostile to the new religion. After that, Muhammad bade Muslims to face Mecca rather than Jerusalem when they prayed, now five times a day, rather than three times.

Earlier, Mohammad had been very impressed with the more ancient and established Jewish religion of Yahweh, and had commanded Muslims to pray three times a day, facing Jerusalem, as was the custom of the Jews.

He had allowed intermarriage and proclaimed a fast day to coincide with the Jewish Day of Atonement. He had accepted the earlier prophets, Abraham, Noah, Moses and Jesus. However, he did not believe Jesus was the "Son of God" (Allah) nor that he had been resurrected.

Note that Mohammad also considered Jews and Christians to be "People of the Book", the Old Testament.

PALESTINIANS VERSUS ISRAEL

[REFS 43-58]

For centuries, the search for a national home for the Jewish people, in Palestine or elsewhere, had been debated, by both orthodox and reformed Jewish factions. This was referred to as the *Zionist Movement* (43)(44)(45).

One of the most extreme Zionists was Rabbi Meir Kahane, who is quoted as saying, among other things, that "There is only one message: God wanted us to come to this country (America) to create a Jewish state." He was assassinated in New York in 1990 (46).

Perhaps second only to the Christian Crusades for the Holy Land, the final insult to Islam could have been the British colonial gift of Palestine to the Jews (1948 AD/CE).

A United Nations resolution in 1947 had partitioned Palestine, creating Israel, and the British troops withdrew.

Jewish people had been settling there since 1820 and then "poured in" during Hitler's persecution. The crisis probably dates to the League of Nations *mandate* of Palestine to Great Britain in 1922.

In 1917 the British had declared their intent to establish a national home for the Jewish people within Palestine, but that nothing should prejudice the civil and religious rights of non-Jewish communities in Palestine. This was referred to as the Balfour Declaration. British troops then occupied Palestine. In 1919, surveys indicated strong opposition to an independent Israel. The British ignored this.

Around 1920, of a total Palestinian population of 700,000, about 80% were Muslim and 20% Christians and Jews. Before WW II there were 1.3 million Palestinians. After the Nazis came to power in Germany in 1933, over 60,000 Jews fled to Palestine. An Arab revolt, opposing possible Jewish domination, lasted from 1936-1939. After that point, Britain limited Jewish immigration and purchase of land (47).

After 1945 some 100,000 survivors of Nazi death camps attempted to flee to Palestine, despite British opposition.

In 1947 the United Nations voted for a plan to divide Palestine into a Jewish and an Arab state with Jerusalem under international administration. The United Nations plan was accepted by the Jews but rejected by the Arabs. Jews and Arabs prepared for war. At that time there were 610,000 Jews in Palestine (32% of the population (48)). When the British withdrew on May 14, 1948, Israel proclaimed itself an independent state. On the following day five Arab armies attacked Israel – around 18,000 troops from Egypt, Transjordan, Iraq, Syria and Lebanon. Israeli troops numbered around 62,000.Israeli dead were over 6,000. Arabs lost over 2,000 from regular armies plus irregulars. A cease-fire was reached on January 7, 1949.

Israel had increased its territory by one-half. Jordan annexed adjacent Arab territory. Egypt occupied coastal areas including Gaza. Later, in 1949, Israel allowed 150,000 Arab refugees to return, mostly to reunite with their families. Some 400,000 Palestinian Arabs who had fled Israel were settled in refugee camps, remaining the largest in the world (49)(50).

An armed truce between 1949 and 1956, overseen by the United Nations was punctuated by raids and reprisals, with the U.S., Britain and France siding with Israel and the USSR supporting Arab demands. Detailed accounts of these events are available elsewhere (51)(52).

In October 1956, after Egypt took over the Suez Canal, Israel assaulted Egypt's Sinai peninsula, followed by a British-French occupation of the Canal. A cease-fire left Israel occupying Gaza and Sharm-el-Sheikh. After "strong pressure" from the U.S., USSR and the U.N., Israel left the Sinai in November 1956 and Gaza in 1957. This was only after Israel's continued occupation of Sharm (Israel's only access to the Indian Ocean) was guaranteed.

Beginning in 1963 the Syrian Army shelled Israeli villages from the Golan Heights. The next Israeli-Arab war (the "Six-Day War) began in 1967. In May, Egypt blockaded the Israeli port of

Eilat on the Gulf of Tiran. On June 5th, Israel attacked Egypt and Syria. Jordan attacked Israel, which then occupied Gaza, the Sinai Peninsula, Syria's Golan Heights, the West Bank and the Jordanian sector E of Jerusalem. Arab guerilla incursions continued, mostly from Jordan.

Still another Islamic-Israeli war began on a Jewish holy day, October 6, 1973.It was called the Yom Kipper War. The war ended with the defeat of Egypt and Syria, following which a U.N. Security Council brokered cease-fire became effective on October 23, 1973.

In 1979, Egypt and Israel signed a peace treaty at Camp David, with the help of President Jimmy Carter. Egypt gave full recognition to Israel. Trade relations between the two countries were resumed and the Sinai was returned to Egypt.

On June 6,1982, Israel invaded Lebanon for the second time to eliminate Palestine Liberation Organization (PLO) bases used in raids on Northern Israel. Israeli troops surrounded West Beirut. They withdrew by 1985 after the PLO left. Excellent summaries and updates are available (53).

The wholesale slaughter of Jewish and Palestinian civilians continued in 2005 at the time of this initial writing. Especially since the 1993 Oslo Peace Accord, the deadlock could have been attributed mostly to two individuals, Ariel Sharon, Israeli Prime Minister, with his modern Army, and Yassir Arafat,PLO Chairman, with his young stone-throwing suicide bombers. The latter protested the continued *illegal* occupation and settlement of Palestinian territories taken by the Israelis in the 1948 War.

Between 1993 and 1995, there were agreements between Israel and the PLO for mutual recognition, transition to Palestinian self-rule in Gaza and Jericho, but Islamic terrorist attacks continued. In early 1996 there were more suicide bombings and rocket attacks from Lebanon by Shiite Muslims. Israel blockaded the port of Beirut and attacked targets in South Lebanon.

For years now, the United Nations has had the Israeli-Palestinian conflict high on its agenda. No real progress has been evident (54).

Neither Arial Shannon, Israel's Prime Minister at that time, nor Yassir Arafat, the PLO Chairman, or their many supporters, had seemed willing to accept anything but complete mutual extermination. Since Arafat's death in late 2004, the future has continued to be very uncertain.International law references (occupied territories) are available to help clarify the ongoing internal conflict between Israel and occupied Palestine (55)(56)(57)(58).

JEWISH WARRIORS

[REFS 59-62]

Since 1843, the B'nai B'rith International has defended the Jewish religious community worldwide, working with the United Nations and heads of different governments.

As a religion which does not seemingly attempt to convert "heathens," it is also not as clearly competing as "Rulers of the World," unless in relation to international banking, e.g., Rothschilds.(See *International Banks* under *Shadow Governments* (59).

Historically, Jewish bankers have helped finance wars. The first truly international banks were probably those of the Jewish family Rothschild. However, the Knights Templar in Europe, between 1150-1300 AD/CE, could be considered to have been earlier multinational bankers.

For many centuries Jews have been persecuted, massacred and driven from countries they called "home," especially since the major dispersions of diasporas from Jerusalem, when Nebuchadnezzar plundered and then destroyed the city (597-587 and 581 BC/BCE).

After that, Jews "scattered upon the faces of the earth." Another dispersion did occur in 66 AD/CE as a revolt against Roman rule failed. Jerusalem was captured and the temple destroyed.

By 538 BC/BCE the Jews were told they were to be messengers to the world as God's chosen people.

The reasons for persecution (or excuses) have sometimes been their refusal to "convert" to a more dominant and militant religion, such as Islam or Catholicism. Also, it may have been related to envy or resentment of the "closed" nature of Jewish communities. This would especially be true it, as "God's chosen people," this reflected a sort of arrogance (60)(61).

Examples of Jewish "warriors" should include those who struggled to survive Hitler's and Stalin's death camps. Also, when they became homeless refugees in 1492 AD/CE, after being expelled from Catholic Spain, they fled to the Ottoman Empire, to the "Holy Land" at Safed in Galilee, only to be assaulted by Arabs.

Again, in 1882 AD/CE, after the Russian Tsar Alexander II was assassinated, anti-Semitism escalated and some "ardent young socialists such as Ben Gurion", failing in their efforts to actualize the theories of Karl Marx in Russia, fled to Palestine. Under Tsar Alexander III, Catholics, Protestants and, especially Jews, were attacked (c.1881 AD/CE) (62).

In modern Israel, the Mossad is the intelligence agency responsible for national security and operates much like the CIA in the U.S. [See http;//en.wikipedia.org/wiki/Mossad]

THE CONVIVENCIAS

[REFS 63-66]

Convivencia means *living together* – in the Spanish language. In all fairness to history, there were other places and times where people of different cultures and religions lived together in peace and prospered, much like the *Two Greatest Convivencias* in Cordoba, Spain, and Alexandria in Egypt.

For example, this condition was surely evident at times in Macedonian Constantinople, the *New Rome*, "center of the civilized Christian world," from its dedication in 330 AD/CE by the first Christian emperor, Constantine I until 1453, when it fell to the Ottoman Turks. It survived its alienation from the Church in Rome and almost endless sieges, including marauding Crusaders in 1203. It was probably at its zenith under Constantine VII, between 944-959 AD/CE. The population approached a million, ten times that of any Western capital, with Muslims, Jews, Christians, Syrians, Armenians, Greeks, Russians, Bulgars, Italians, Germans, Normans, Anglo-Saxons, "rubbing shoulders" in the capital of the Byzantine Empire, the "greatest commercial power in the world." The University of Constantinople had faculties of law, medicine, philosophy, mathematics, and astronomy (63)(64). Another example might be Baghdad in the age of the Abbasid caliphs, the heart of the Islamic Empire, "the greatest of all cities," between 762-1258 AD/CE (65)(66).

However, there were at least two "shining moments" in time when differing religious cultures flourished together: Alexandria, Egypt and Cordoba, Spain. Alexandria's history began about 322 BC/BCE. Cordoba's history could be said to begin with the Visigoths in Spain in 587 AD/CE, when their king converted to Catholicism or, perhaps, with the first invasion by Arab Muslims and Berbers from North Africa in 711 AD/CE.

This lengthy "tale of two cities" extends over hundreds of years. Detailed and colorful descriptions of these cities, so widely separated in time, are available in *APPENDIX C: The Two Great Convivencias – Alexandria and Cordoba*.

REFERENCES AND NOTATIONS: CHAPTER FIVE

THE HOLY WARRIORS

1) Garraty and Gay. *The Columbia History of the World*,1972. [See Chapter 35]
2) Meyers,Neil. Holy Wars Across Time. www.buzzle.com/editorials/text10-30-2002-29238.asp.
3) Srayer,Joseph (Editor).*Dictionary of the Middle Ages.* New York. Charles Scribner,1984.[Volume 4]
4) Paris, Edmund. *The Secret History of the Jesuits.* 1975. Ontario, CA. Chick Publications.[See pages 155-162 and 177-182]

CATHOLICS VERSUS JEWS

5) Toland,John. *Adolf Hitler.*New York.Ballantine Books, 1976. [See pages 249,364,431, 548]
6) Bromberg,Norbert,M.D.;Small,Verna. *Hitler's Psychopath- ology.*New York.International Universities Press,1983.
7) Carroll,James. *Constantine's Sword: The Church and the Jews.* Boston & New York. Houghton Mifflin, 2001.
8) Paris,Edmund.*The Secret History of the Jesuits:Translated from the French1975*.Ontario,California.ChickPublications.

SOCIETY OF JESUS (THE JESUITS)

9) Paris,Edmund. *The Secret History of the Jesuits.* 1975. Ontario, CA. Chick Publications.
10) Saussy,F.Tupper. *Rulers of Evil.* New York. HarperCollins, 1999.
11) Picknett,Lynn; Prince,Clive. *The Templar Revelation: Secret Guardians of the True Identity of Christ.* New York. Touchstone, 1997. [See Chapter Four]
12) Wright,Jonathan. *God's Soldiers – A History of the Jesuits.* New York. Doubleday, 2004.
13) Picknett, Prince,Clive. *The Templar Revelation: Secret Guardians of the True Identity of Christ.* New York. Touchstone, 1997.[See Chapter Four: *Heartland of Heresy*]
14) Saussy,F.Tupper. *Rulers of Evil.* New York. HarperCollins, 1999. [See pages 83 and 104]
15) Saussy,F.Tupper. *Rulers of Evil.* New York. HarperCollins, 1999. [Chapters 7-14 and Pages 296-297]
16) Martin,Malachi. *The Jesuits: The Society of Jesus and the Betrayal of the Roman Church.* Touchstone/Simon and Schuster,1987.
17) TIME Almanac-2006.[Pages 367-369:*Roman Catholic Pontiffs* - St. Peter (c.42 AD/CE) to Benedict XVI (2005 AD/CE]

18) Paris,Edmund. *The Secret History of the Jesuits.* 1975. Ontario, CA. Chick Publications. [See page 24]
19) Paris,Edmund. *The Secret History of the Jesuits.* 1975. Ontario, CA. Chick Publications.[Part IV, Chapter 8: *The Jesuits,General Boulanger and the Dreyfus Affair*]
20) Paris, Edmund. *The Secret History of the Jesuits.* 1975. Ontario, CA. Chick Publications. [Part IV, Chapter 8:*the Jesuits, General Boulanger and the Dreyfus Affair*]

ISLAMIC MUJAHIDEEN

21) Esposito,John (Editor).*The Oxford History of Islam.* New York. Oxford University Press, 1999. [NOTATION: See pages 102,658]
22) Rashid,Ahmed. *Jihad: The Rise of Militant Islam in Central Asia.* Yale University Press, 2002. [See page 6]
23) Bergen,Peter. *Holy War, Inc: Inside the Secret World of Osama bin Laden.* [Page 89]
24) Friedman,Thomas. *The ABC's of hating the "infidels."* New York. Times News Service, June 4, 2004.
25) TIME Almanac – 2006. [See pages 796-799: *Israel, History.*]
26) Israeli,Raphael (Translator).Jerusalem, Israel. Harry Truman Research Center. The Hebrew Institute.[This lengthy *Hamas Charter* was available over the Internet.]
27) Lewis, Bernard. *What Went Wrong? The Clash Between Islam and Modernity in the Middle East.* New York. Perennial- HarperCollins. Oxford University Press, 2002 [See Waqf on pages 110-111]
28) www.maintou.com/bcof/bcofw040.htm AND www.library.nps.navy.mil/home/tgp/qaida.htm
29) Bergen,Peter, 2002. [Same reference as (23)]
30) Emerson,Steven. *American Jihad: The Terrorists Living Among Us* .New York and London. The Free Press, 2002.
31) Baer,Robert. *See No Evil: The CIA's War on Terrorism.* New York. Crown Publishers, 2002.
32) *The 9/11 Commission Report – Authorized First Edition,* 2004. [*Chapter Two: Foundation of the New Terrorism.* Probably the most concise and accurate summary to date.]
33) Bodansky,Yossef. *Bin Laden: The Man Who Declared War on America.* New York. Random House,1999.
34) Friedman,Thomas. *From Beirut to Jerusalem.* New York. Anchor Books, 1995.
35) World Almanac-2001. [See *Israel*- pages 800-802.]

36) Sanger,Peter (General Editor). *The Encyclopedia of World History: Ancient, Medieval and Modern – Chronologically Arranged.* New York. Houghton Mifflin Company, 2001.

37) Ayton-Shenker,Diana (Editor).*Global Agenda: Issues Before the 56th General Assembly of the United Nations.* New York and Oxford. Roman and Littlefield, 2002. [Notation: See *Israel,* pages 50,176-182; Taliban, pages 50,201; *Terrorists,* pages 169,282; *Iraq,* pages 42,173- 176,291-292; *Jihad* against Christians in Indonesia, page 213]

MUSLIMS VERSUS JEWS

38) Garraty and Gay. *Columbia History of the World,* 1972. New York, London. Harper & Row,1972.[See chapters 8, 22,25]

39) Armstrong,Karen. *A History of God: The 4,000-Year Quest of Judaism, Christianity and Islam.* New York. Ballantine Books,1993. [See Chapter 2: One God]

40) Garraty and Gay. *Columbia History of the World,* 1972. New York & London. Harper & Row,1972. [See chapters 8, 22,25]

41) Peters,F.E. *Children of Abraham: Judaism, Christianity, Islam.* Princeton, NJ. Princeton University Press, 1982.

42) Feiler,Bruce. *Abraham: A Journey to the Heart of Three Faiths.* New York. *HarperCollinsPublishing,*2002.

PALESTINIANS VERSUS ISRAEL

43) Armstrong,Karen. *A History of God,*1993. [See page 372]

44) Catholic Encyclopedia:Zionists. Sources cited in 1899, 1902, and 1905. [Available at: www.newadvent.org/cathen.15760c.htm]

45) Weber,Mark. Zionism and the Third Reich. [www.ihr.org/jhr/v13n4p29_Weber.html]

46) Armstrong,Karen. *A History of God,* 1993.[See page 391]

47) Bauer,Yehudi. *From Diplomacy to Resistance: A History of Jewish Palestine.* Philadelphia, PA. Jewish Publication Society,1970.

48) Stearn,Peter (General Editor).*The Encyclopedia of World History – Sixth Edition.* New York. Houghton Mifflin,2001. [See page 966.]

49) TIME Almanac–2006.Time,Inc. Boston, 2005. [See Israel, pages 796-799]

50) Caner,Ergun; Caner,Emir. *Unveiling Islam.* Grand Rapids, MI. Kregel Publishers, 2002.

51) Garraty and Gay. *Columbia History of the World,* 1972. New York, London. Harper & Row, 1972.[Pages 1049 and 1101]

52) Stearns,Peter (General Editor). *The Encyclopedia of World History. Sixth*

Edition. New York. Houghton Mifflin Company, 2001.[See pages 532, 762-764, 966-968,979-982]

53) Friedman,Thomas. *From Beirut to Jerusalem.* New York. Anchor Books,1995.

54) Ayton-Shenker,Diana; Tessitor,John (Editors).*Global Agenda: Issues Before the 56th General Assembly of the United Nations.* New York and Oxford. Roman and Littlefield,2002. [See pages 176-182]

55) Von Glahn,Gerhard. *Law Among Nations: An Introduction to Public International Law – 2nd Second.* New York. Macmillan Publishing, 1970.[See Chapter 31:*Belligerent Occupation*] [Also see *War Crimes and Crimes Against Humanity,* page 702]

56) Gutman,Roy;Rieff,David. *Crimes of War: What the Public Should Know.* New York. W.W. Norton, 1999.

57) Meron,Theodor. *War Crimes Come of Age.* Oxford. Clarendon Press, 1998.[See Chapter XIII: *International Criminalization of Internal Atrocities*]

58) Carter,Jimmy. *Palestine: Peace Not Apartheid.* New York and London. Simon and Schuster, 2006.

JEWISH WARRIORS

59) Ferguson,Niall. *The House of Rothschild: Money's Prophets, 1798-1848.* Paperback ,2000.[See: Ferguson,Niall. *The Cash Nexus: Money and Power in the Modern World, 1700- 2000.*New York. Basic Books,2001. Also See pages 157-168, 304-306,386 (WHO) and 261 (globalization)]

60) Armstrong,Karen. *A History of God, 1993.* [See page 1]

61) Kent,Charles Foster. A History of the Jewish People during the Babylonian, Persian and Greek Periods. New York. Charles Scribner's Sons,1906. [See at www.questia.com]

62) Stearn,Peter (General Editor).*The Encyclopedia of World History.* New York. Houghton Mifflin, 2001.[See page 510]

THE CONVIVENCIAS

63) Vrettos,Theodore. *Alexandria: City of the Western Mind.* New York and London. The Free Press, 2001.

64) Garraty and Gay (Editors). *Columbia History of the World.* New York, London. Harper & Row, 1972.

65) Lewis,Brenda (General Editor). *Great Civilizations.* Bath,UK. Parragon Publishing, 2002.[See *Byzantine Empire,* page 254]

66) Goodwin,Jason. *The Glory That Was Baghdad.* Palm Coast, FL. The Wilson Quarterly- Spring 2003.

CHAPTER SIX

RELIGION VERSUS SCIENCE

RELIGION VERSUS SCIENCE OVERVIEW

[REFS 1-15]

In contrast to the methods of science, a religion is here defined as any belief system based on *wishful or fearful thinking* – and not from any reproducible and verifiable evidence in the natural world. Usually, this includes the *isms*, such as *Judaism, Catholicism* and *Muhammadism* . Historically, doubts or critical examination have been discouraged or forbidden, sometimes on pain of death, by some very authoritative *faith* organizations. Horrendous *crimes against humanity* have been committed under the cloak of religion and attributed to the *will of God or Allah* – as proclaimed by some self-appointed prophet or religious leader. It is really about power. This should be most apparent where someone attempts, allegedly, to *save the world by conquering or colonizing,* whether or not it is in the name of some God (1)(2)(3)(4)(5)(6)(7).

In the United States in recent years there has been considerable controversy over embryonic stem cell research – with the *religious right* versus *science* (8)(9).

This topic of religion versus science also relates to the role of religion from the French Revolution in 1776 to the United States in

1887 and into America in the twentieth century (10)(11)(12)(13)(14).

A recent New York Times *Best Seller*, extensively researched and written by *an acclaimed author of many books*, could be the most thorough coverage of the entire debate over the reality of one or any Gods (15).

ORIGINS OF RELIGION

[REFS 16-31]

The earliest hint of human belief in a *life-after–death*, was probably when the Neanderthals (homo sapiens *Neanderthalensis*) began burying their dead rather than eating them or exposing them to the elements. Neanderthal burial sites in the Northern hemisphere have been dated to around 50,000 years ago in Israel and south-western France. Egyptian pyramid tombs date only to about 2700 BC/BCE (16)(17).

The next stage, involving the burying of "travel items" with the dead, became evident with the appearance of a "truly modern man"(Homo sapiens *sapiens*).This "modern man" appeared in Africa somewhere between 90,000 and 125,000 years ago (18)(19)(20).

Archaeological research has demonstrated unexplained similarities among people completely separated by different continents and great distances throughout the ancient world in regard to burial and sacrificial practices, such as the building of temples, including pyramids, and the worship of various gods. Belief in an after-life and, at times, ancestor worship, have both been apparent (21)(22)(23).

Human settlements in China have been dated to as early as 5,000 BC/BCE. North American mound-builders' burial sites date to around 1500 BC/BCE, and the temples in Meso-america to around 1000 BC/BCE. This is compatible with the theory that the first humans came to America over the frozen Bering Strait before the rapid warming which started about 15,000 years ago (24).

Other than the Egyptian pyramid tombs, one of the most spectacular ancient funereal monuments are the 8,000 to 10,000 terracotta warriors guarding the tomb of the first emperor of a united China, Qin Shi Huangdi (259-210 BC/BCE), located at Xi'an.

Much earlier, native American mound-builders, as early as 3400 BC/BCE, were burying their dead in tombs or, perhaps, rudimentary temples (25).

The first consciously-considered belief in some kind of personal *life-after-death* must have had to await a dramatic evolutionary change in the human brain. This *life-after-death* belief is most pervasive and persistent and is the core of many religions. It must also relate to the extremely powerful survival instinct seen in all animals.

This brain change became evident some 50,000 years ago with what has been called the *dawn of human culture*, manifested by dramatic changes in *behavior*, but not in the size of the brain. For some 200,000 years the brains of Homo sapiens and their ancestors have remained about the same size. These behavioral changes included a new kind of "self-awareness", imagination, inventiveness and use of symbols. With this self-awareness came the ability to consider personal past and future. This momentous change was probably the result of a mutation in DNA (26)(27)(28).

Self-appointed prophets, seers, shamans, fortune-tellers or witches have, for millennia, deceived multitudes by claiming to speak for some invisible human-like supernatural being, god or goddess.

These imaginary beings were given credit or blame for "miracles" and catastrophes. It naturally followed that these mysterious events were a result of obedience or disobedience of the rules or laws provided by the person claiming to communicate with the god or gods.

It would seem reasonable, in the light of historical and current knowledge of human behavior, to suspect that some useful basic rules relating to survival could have been the result of thousands of years of humans interacting with each other, however slow to learn from experience. A *GOD* might not be necessary, except to give the alleged *messenger* great power.

The list of *Sins* (disobedience to some supreme invisible ruler or rulers of the Universe, including "taking the life or the wife of another, false accusation, failing to give water to one in need") must surely reflect *learning,* where human *experience* has interacted with some basic human *instincts,* rather than *Laws* given by some imaginary *God* – or angel.

The list of rewards and punishments, provided by men, have varied widely through the ages. Punishment by the gods, in many cases, consisted of plagues and natural disasters. In the most extreme cases, punishment has been carried out by those who claimed to speak for a god, carrying out deliberate torture and execution of the "sinner or disbeliever". These punishments have been described in horrifying detail in a number of writings. These are not for the "faint-hearted" (29)(30)(31) .

Belief in another life after one's *physical* death has surely brought comfort to multitudes over thousands of years but, tragically, has also been used by many self-appointed *prophets or saviors* to control or destroy millions of *non-believers.*

EARLIEST RELIGIONS

[REFS 32-36]

Even before the end of the fourth and most recent *Ice Age,* around 10,000 years ago, there were many gods. A single supreme god was little in evidence (32)(33) .

In the following text, where possible, dates are given for events, to give a feeling of *connectedness* and to encourage further research by a reader.

Between 10,000 and 4,000 years ago, hunting economies were replaced by agricultural communities in Mesopotamia and later in Egypt. Some sources place the migration of "Semitic hordes" from the Arabian desert to lower Mesopotamia (Sumeria) as early as 5,000 BC/BCE. Most moved beyond Sumer, further North, up the Euphrates River. Another source describes " a constant struggle" between the older Sumerian (Kengi) cities in the South and their Northern Semitic (Kish) kinsmen.Sumerian pottery and a "white temple" in Uruk have been dated at around 4,000 BC/BCE. This was the Neolithic or New Stone Age. The earliest copper and bronze tools have been dated at 2,500 BC/BCE.

The many gods of Sumeria-Babylon had a significant effect on the ancient Israelites and the newer religion of Yahweh in Canaan (34)(35)(36).

Around 4000 BC/BCE, in what is now Iraq, the Sumarians established one of the first great cultures of the known world. Major cities listed were Ur, Erech and Kish. Babylon was also settled by Sumarians. Later, Semitic Akkadians invaded, adopting the language and culture of Sumaria.

Records of the many Sumarian kings began with Mebaragesi of Kish, about 2700 BC/BCE. Around 2500 BC/BCE, the Akkadians of Sumaria developed a written script of 550 symbols. Clay tablets date before 2000 BC/BCE. In Egypt, the Great Pyramid of Khufu had been completed by 2500 BC/BCE.

Between 1950-2000 BC/BCE, Semetic Amorites conquered Sumaria and made Babylon their capital. In 1800 BC/BCE, Babylonians and Syrians still identified themselves as Sumerians. (*Semites* may describe all groups speaking any language of the *Great Family,* which included Arabic, Hebrew, Phoenician and Egyptian).

An Amorite, Hammurabi of Babylon (1792-1750 BC/BCE), built

an empire from the north Euphrates to the Persian Gulf and enlarged upon older Sumarian laws. These included the *Lex Talionis,* an eye for an eye and a tooth for a tooth. In Mesopotamia there were four main gods: Anu, Enil, Enki(Ea)and Ninhursag.

Between 1700-1600 BC/BCE, Aryans, from present Iran, invaded and destroyed the Indus Valley civilization. They also found "a multitude of gods." In Egypt (c.1786 BC/BCE), Semitic Syrians ended the Middle Kingdom and the 12th Dynasty (1991-1785 BC/BCE).

Sometime after 1500 BC/BCE, Assyrians had settled in nearby Ashur. The Assyrian Empire is recorded as 935-612 BC/BCE. Sometime between 800-700 BC/BCE Assyrians conquered Babylon. Sumer became part of Babylon between 600-500 BC/BCE.

SCIENCE AND RELIGION

[REFS 37-44]

For several millennia of recorded history, it has been shown that – under the influence of powerful emotions – reason, logic and judgment often disappear. There is no more powerful emotion than the fear of death (or perhaps, torture). Humans are the only animal capable of anticipating their personal death as inevitable but, also, of denying its permanence.

No *Holy Scripture* nor revelation given by some god or angel to some self-proclaimed prophet has been responsible for the almost miraculous discoveries of Science. These certainly include the conquest of many devastating diseases and the ability to land men on the moon. Disease pandemics, as well as other natural disasters were, in the not-so-distant past, attributed to some god's wrath and punishment for "sins."

One definition of *atheist* could be: "A person not believing in any supernatural, invisible or imaginary *supreme beings,* created in the image of Man by men." Nevertheless, an atheist may be supremely *religious* in the sense of having the primitive's and the child's tremendous awe of nature, together with their curiosity and *humbleness.* The latter is the mark of any true scientist. *Science* has a dictionary definition of:

1) A branch of knowledge or study dealing with a body of facts or truths systematically arranged and showing the operation of general (natural) laws.
2) Systematic knowledge of the physical or material world.

A more comprehensive description of the concept of *Science* may be found in a recent textbook (37).

An early classic description of the "nature of scientific enquiry" was by Robert Grosseteste (c. 1169-1253 AD/CE) in England. He availed himself of the many translations from the Arabic and Greek sciences. Briefly, he pointed out that the aim of science was to discover the reasons or causes for natural phenomena. Next, to be followed analysis for underlying principles. Next, the formation of a hypothesis or theory which must be tested and proven or disapproved (38).

More current and comprehensive presentations of the past discoveries of science and the continuing search for under-standing the mysteries of life and the universe are available in the following references (39)(40).

The conflicts and the agreements between religion and science are examined at length by 31 scientists in a recent book (41).

Another in-dept examination of the psychological roots of religious beliefs could be especially informative (42).

The great extend of Arabic contributions to the sciences during the "Dark Ages" of Medieval (Catholic) Europe may not be commonly known today (43).

Medieval Europe and the "Dark Ages" may be considered as beginning with Constantine (c.312 AD/CE) and with the gradual decline of the Roman emperors. The ascendancy of Catholicism in Europe and Islam in Africa and Asia followed, especially after 600 AD/CE.

For the first thousand years after creation of the Koran by Muhammad ibn Abd Allah, the last prophet (c.570-632 AD/CE), "the world of Islam was in the forefront of human civilization and achievement."

Between 800-900 AD/CE, Arabs became more aware of Greek science and philosophy, with a host of translations from Greek to Arabic. A new type of Arab thinker emerged, called *Falsafah* (philosopher), dedicated to living according to laws governing the cosmos (44).

However, by 1900 AD/CE, "it was abundantly clear in the Middle East and indeed all over the lands of Islam that things had indeed gone badly wrong." Europeans "began to win victory after victory, first on the battlefield, then in the market place." "Was this, in a sense, the victory of science over religion? This is discussed in exquisite detail elsewhere (43).

In the present author's opinion, the greatest contribution of synagogues, churches an mosques to human cultural evolution is in providing an extended-family to members in need and encouraging responsible world citizenship, which must include religious tolerance for the beliefs of others. It is not necessary for organized religions to promote the hoax of "heaven or hell" after one's death. By this time in history, humans should be capable of realizing, from their own experience, the difference between "right and wrong" in terms of behaviors which assure the survival and continued evolution of their species. You do not kill someone for not believing your own possible delusions. The only thing worse than death is being deceived about it.

Human Nature, instinctive in origin, has not changed in 100,000 years and cannot change much in another 100,000. To use the example of a *super-computer*, the three-pound human brain is the hardware. Human survival requires a change in their brain *software – information*. This is through learning the facts about *human nature* – and the universe.

This requires the study of human history and the methods of science.

THE PHILOSOPHERS

A discussion of Religion and Science would seem incomplete without considering the *Philosophers* in history.

Perhaps a brief, yet cogent, definition of *Philosophy:*

"The study of *knowledge.*"

The world-view of philosophers has changed with new knowledge of the *natural* world and *human nature.*

This seemed to begin with the "great creativity" in Greece between 750-250 BC/BCE – before Rome's dominance began around 500 BC/BCE. [One of the best introduction to the greatest philosophers can be found in Will Durant's *The Story of Philosophy: The lives and opinions of the world's greatest Philosophers from Plato to John Dewey.* New York. Simon & Schuster,1926-1961.]

REFERENCES AND NOTATIONS: CHAPTER SIX

RELIGION VERSUS SCIENCE OVERVIEW

1) Armstong, Karen. *A History of God, 1993.*[See pages 258 and 196-204]
2) Bauer, Yehudi. *From Diplomacy to Resistance: A History of Jewish Palestine,1930-1945.*Philadelphia.Jewish Publication Society, 1970.
3) Bodansky, Yossef. *Bin Laden: The Man Who Declared War on America.* New York and London. Forum /Crown Publishing, 1999.
4) Carroll, James. *Constantine's Sword: The Church and the Jews.* New York. Houghton Mifflin Company, 2001.[See pages 237-312 and pages 313-400]
5) Emerson, Steven. *American Jihad: The Terrorists living Among Us.* New York and London. The Free Press, 2002.
6) Kimball, Charles. *When Religion Becomes Evil: Five Warning Signs.* HarperSanFrancisco, 2002.
7) Rashid, Ahmed. *Jihad: The Rise of Militant Islam in Central Asia.* New Haven, CT. Yale University, 2003.
8) Money, Chris. *The Republican War on Science* .New York. Basic Books, 2005.[See Chapter 12:*Stemming Research – and Chapter 11: Creation Science.*]
9) Lowry,Rich. *Another triumph of science over politics: Stem cell research.* Portland, OR. The Oregonian, November 27, 2007.
10) Zamoyski, Adam. *Holy Madness: Romantics, Patriots and Revolutionaries,1776-1871.* New York. Penguin Books, 1999. [See Chapter 4: *False Gods*; Chapter 6: Holy War; Chapter22: *After Life.* NOTATION: A vivid and detailed historical Essay]
11) Phillips, Kevin. *American Theocracy: The Peril and Politics of Radical Religion, Oil, and Borrowed Money in the 21st Century* .New York. Penguin Group,2006.
12) Kurtz,Paul (Editor).*Science and Religion: Are They Compatible?* Amherst, NY. Prometheus Books, 2003. 125
13) Joshi, S.T.(Editor). *Atheism.* Amherst, NY. Prometheus Books,2000. [NOTATION: Thirty-one author's essays on deism, agnosticism and atheism, from 60 BC/BCE (Lucretius) to 1995 (Carl Sagan)]
14) Morain, Lloyd and Mary. *Humanism as the Next Step.* Amherst, NY. Humanist Press, 2000.
15) Dawkins, Richard. *The God Delusion.* New York. Houghton Mifflin, 2006. [A must-read for anyone still in doubt about man-made gods, resurrections or life-after-death]

ORIGINS OF RELIGION

16) Garraty & Gay (Editors).*Columbia History of the World*. New York and London. Harper & Row,1972.[See *Human Evolution,* page 45]
17) Owen,Francis. *The Germanic Peoples*. New York. Barnes &Noble, 1960. [See *Religion*, pages 183-209]
18) Stearns,Peter. *The Encyclopedia of World History –Sixth Edition*. New York. Houghton Mifflin, 2001.[See *Human Origins,* pages 5-8]
19) Garraty & Gay (Editors).*Columbia History of the World*. New York and London. Harper & Row,1972. [See Chapter 10: *Early China and Egyptian Religions,* pages 69-73, 91-94]
20) *Past Worlds: Atlas of Archaelogy*. HarperCollins/Borders Press, 2001.[See *Burials,* page 34 and *Rituals,* pages 48-49]
21) Stearns,Peter. *The Encyclopedia of World History –Sixth Edition*. New York. Houghton Mifflin, 2001. [See *Egypt*, page 28; *Olmecs and Mayans,* page 19]
22) Scarre,Chris. *Exploring Prehistoric Europe*. Oxford University Press, 1998.[See *Stonehenge (2950-1600 BC/BCE);Hochdorf (c. 550 BC/BCE)*] *Edition*. New York. Houghton Mifflin, 2001. [See page 11]
24) www.geocities.com/Athens/Forum/6558/mound1.html
25) Klein,Richard;Edgar,Blake. *Dawn of Human Culture: A Bold New Theory on What Sparked the "Big Bang" of Human Consciousness*. New York. John Wiley & Sons, 2002. [See pages 144-146 and 270-272]
26) Garraty & Gay (Editors).*Columbia History of the World*. New York and London. Harper & Row,1972.[See Chapter 8: *Gods and Men ; Prehistoric Beliefs*]
27) Stanley,Stanley.Earth *and Life Through Time – Second Edition,*1989.[See *Genes,DNA and Chromosomes,* pages 140-142]
28) Streeter,Michael. *Witchcraft:A Secret History.* Hauppauge, NY. Barrons Educational Series, 2002.[See pages 97-103, *Trial By Torture and Ordeal*]
29) Farrington,Karen. *History of of Punishment and Torture*. Octopus Publishing,1996 and 2002.
30) Swain,John.*The Pleasures of the Torture Chamber.* New York. Dorset Press, 1931.[Also see *Anarchists* and the *Fine Art of Torture.* Guardian Newspapers, Ltd., January 2003]
31) Ridley,Jasper. *The Freemasons,* 2001.[Chapters 2 (Heretics) and 5 (Pope's Bull)]

EARLIEST RELIGIONS

32) Armstrong,Karen. *A History of God*,1973.[See Chapter One: *In the Beginning*]
33) Garraty & Gay (Editors).*Columbia History of the World*. New York and London. Harper & Row,1972.[See *Gods and Men*, pages 91-94]
34) Peters,Francis. *The Children of Abraham: Judaism, Christianity, Islam*. Princeton University, 1982.
35) Kent,Charles Foster. *A History of the Jewish People During the Babylonian, Persian and Greek Periods*. New York. Charles Scribner's Sons, 1906.
36) www.publicbookshelf.com

SCIENCE AND RELIGION

37) Kurtz,Paul(Editor).*Science and Religion: Are They Compatible ?* Amherst, NY. Prometheus Books, 2003. [See page 11]
38) Ronan,Colin. *Science: It's History and Development Among the World's Cultures*. New York. Facts on File Publications, 1982. [See pages 252-253]
39) Maddox,John. *What Remains to be Discovered: Mapping the Secrets of the Universe, the Origins of Life, and the Future of the Human Race*. New York. Touchstone/ Simon & Schuster,1998.
40) Bunch, Bryan; Hellemans, Alexander. *The History of Science and Technology, Boston* and New York. Houghton Mifflin Company, 2004.
41) Kurtz,Paul (Editor). *Science and Religion: Are They Compatible?* Amherst, NY. Prometheus Books, 2003.[See Chapter 32]
42) Freud, Sigmund. Translated by W.D. Robson-Scott. *Future of an Illusion*. Garden City, NT. Doubleday and Company, 1927.
43)Lewis, Bernard, *What Went Wrong? The Clash between Islam and Modernity in the Middle East*. New York. Perennial- HarperCollins Publishers and Oxford University Press, 2002.
44) Armstrong, Karen. *A History of God*, 1993.[See Ch.6: *The God of the Philosophers*].[Also see: Durant,Will. *Story of Philosophy*,1961]

CHAPTER SEVEN

SECRET SOCIETIES

THE CANDIDATES

[REFS 1-7]

The *Candidates* as rivals for *Rulers of the world* surely include the world's most influential *secret* societies, such as the International Freemasons, the Skull-and-Bones fraternity at Yale University (an offshoot from the Bavarian Illuminati) and "the three most notorious of modern secret societies."

These *three* are thought to be the Trilateral Commission, the Council on Foreign Relations, and the Bilderberg Group (1)(2)(3).

Why the need for secret societies ? Is it the fear of exposing some vulnerability to enemies, real or imagined ? Is it a fear of discovery by intended victims ? Is it a need to *belong* or to find an *identity* through an exclusive group – street-gangs or the Mafia ?

At any rate, it is obvious that some personal advantage has resulted to members, despite the possibility of some extreme sanctions for *defecting* (4)(5)(6) .

Adam Weishaupt, founder of the *Illuminati* in 1776 Baveria, is quoted as saying "Secrecy gives greater zest to the whole...the slightest observation shows that nothing will so much contribute to increase the zeal of the members as a secret union (7)."

Of current global importance, some better known secret societies include the *Order of Skull and Bones* and the *Free and Accepted Masons.* The *Illuminati* is considered only as a background to these two currently very influential organizations.

ORDER OF SKULL AND BONES

[REFS 8-17]

This secret *Order of Skull and Bones,* a fraternity for sons of the wealthy old-line (Puritan) New England elite, was founded in 1832 at Yale University by William Huntington Russell, was extremely wealthy from his opium smuggling. Earlier, it had been referred to as the *Order* or *Chapter 322* of a German secret society, not otherwise identified. However, the utopian philosophies of Georg Wilhelm Friedrich Hegel (University of Berlin, 1817) were very evident from the writings of Yale faculty after returning from Germany in those times (8)(9).

Similar philosophies were espoused by the secret German *Illuminati* of Adam Weishaupt, Bavaria, 1776, by Jean Jacques Rousseau (1712-1778) and later by Karl Marx and Friedrich Engels in the *Communist Manifesto,*1843. The Illuminati claimed to have had *direct ties* through Masonry to the cult of Isis in Ancient Egypt (10).

A reproduction of a Masonic ceremonial *third degree training board* used in 1825 includes a Masonic ceremonial *third degree tracing board* used in 1825 includes a prominent display of the Skull and Bones. Also note the *skull and crossbones* flag of the Knights Templar's fleet (c. 1300) (11)(12).

In Germany in July 1782, this radical secret society, the *Illuminati* as discussed below, allegedly merged with the Freemasons (13)(14)(15).

Some very wealthy families associated with the Skull-and-Bones at Yale have included: Whitney, Lord, Phelps, Wadsworth, Allen, Bundy, Adams, Stimson, Taft, Gilman, Perkins, Harriman,

Rockefeller, Payne, Davison, Pillsbury, and Weyerhauser.

Eventually some German Jewish banking families were included: Shiff, Warburg, Guggenheim, Meyer. Some later converted to Protestantism. Their sponsors in American included the Rothschilds and the Cecil Rhodes Trust. Some Jewish fraternal societies, such as B'nai and B'rith, were allegedly formed out of the Scottish Rite Freemasonry.

Many members of the Skull-and-Bones have become American Presidents, Supreme Court Justices, and have held high positions in government, (including the CIA or Central Intelligence Agency), in private corporations, and in banks.

There has been much inter-marrying of these prominent families, much as among early European monarchs.

Some other prominent names include: President and Chief Supreme Court Justice William Howard Taft (1909-1913); Senator Robert A. Taft (1938-1950); Senator Prescott Bush, (*father* of George Herbert Walker Bush); and President George Herbert Walker Bush, elected 1988. George H.W. Bush was also a Congressman, U.S. Ambassador to the United Nations, Director of the CIA, and Vice-President under Ronald Reagan) (16)(17).

ILLUMINATI

[REFS 18-23]

The term, *Illuminati*, had been used as early as 1492 in Spain by the Alumbrados. Some form of this secret society had apparently existed for hundreds of years.

In the early 1500s, a young Ignatius Loyola had joined the Illuminati, at that time described as a "secret society of Gnostic fundamentalists who preached that all matter is absolutely and eternally evil." Gnostic has been defined as meaning 'magical knowledge' (18).

By 1556, Loyola was dead at age 65, having served as the first Superior General of the Society of Jesus (1541-1556) and was often referred to as the "Black Pope." His power seemed consolidated at the Council of Trent (1545-1563) (19).

Before 1542, the investigation of heresy was "a local affair." After that date the Jesuits administered the Inquisition, later guided by the Pope's *Directorium Inquisitorium* of 1584. In 1623, the Illuminati were condemned by the Grand Inquisition (20)(21)(22).

In May 1776, Adam Weishaupt, a German law professor and former Jesuit candidate, re-established the illuminati. In 1786, it was "banned" in Bavaria as "diabolical" and dangerous to society. Most candidates were recruited from Masonic lodges. Reportedly, Weishaupt hated the Jesuits and vowed to destroy the Catholic Church and all Christianity. He assumed the "ancient name" of Spartacus (who led the insurrection of slaves against Rome in 73 B.C.). Weishaupt endorsed the "radical" French philosopher, Jean Jacques Rousseau.

In 1771, Weishaupt had been "indoctrinated into Egyptian occultism." His plan was for a single, powerful occult organization to save mankind and rule the world. Jean Jacques Rousseau's philosophy was thought to have "embodied all the principles of later communism" and was much like Plato's *Republic* version of a utopian *Atlantis*. In July 1782 in Germany, Masonry and the Illuminati were secretly joined, with their headquarters in Frankfurt and "under the control of the Rothschild bankers."

The tenets of the Bavarian Illuminati have been summarized as follows (23):

1) Abolition of monarchies and all ordered government.
2) Abolition of private property and inheritances.
3) Abolition of patriotism and nationalism.
4) Abolition of family life and the institution of marriage, and the establishment of communal education of children.
5) Abolition of all religion

THE FREEMASONS

[REFS 24-46]

Freemasonry (Free and Accepted Masons) is reportedly the largest, oldest and best documented of all current global secret societies. Membership in the United States alone is reported as over 2.5 million. This is at least half of the world's members.

Freemasons include the *Order of the Eastern Star* for women, *Order of DeMolay* for boys and *Job's Daughters* or *Rainbow* for Girls. There are many other associated organizations, such as the *Ancient Arabic Order of Nobles of the Mystic Shrine* (the Shriners) open to members with at least 33rd degree status (Scottish Rite or the York Rite, Knights Templar (24)(25) .

While having no acknowledged affiliation with Masonry, the Church of Jesus Christ of Latter-day Saints (Mormon church) was founded in 1830 by Joseph Smith, a self-proclaimed prophet, who claimed to receive revelation through angels, and by several fellow Masons. Mormon Temples and rituals have been similar to those of the Freemasons. Estimated members worldwide: 4.9 million.

In addition, Joseph Smith claimed that the Roman Church, founded on Peter, had long ago lost any claim to authority as the true church of Christ due to corruption. This authority was then restored through him, Joseph Smith, as a new prophet to speak for God. (Martin Luther made similar accusations of corruption against the Roman Church in 1517 AD/CE.)

Allegedly, the oldest known Masonic document was the *Regius* poem, around 1390 AD/CE. Between that year and 1717 AD/CE, Masonic lodges began to accept members who were not in the building trade. In 1717, the first Grand Lodge was formed in England (26).

Masonry has been most evident within the past thousand years, despite alleged origins in secret societies of ancient Egypt.

Between 900 AD/CE and 1600 AD/CE, during the era of cathedral building, many Masonic beliefs and rituals were evident in the European stonemason guilds (27) .

In the mid-12th century, Knights Templar called each other *frère macon* or brother Mason, anglicized to *Freemason.* Many writers provide evidence of Freemasonry originating with the Knights Templar.

During the 1700s, Freemasonry had literally "exploded" in the American colonies and was very involved in the Revolution and founding of the United States. The American Revolution (1776 AD/CE) allegedly was planned and carried out largely by Freemasons on both sides of the Atlantic. America was to become the "New Atlantis" as described by Plato (28).

The French Revolution (1787-1789 AD/CE) was allegedly inspired by French Masons and German Illuminati. In 1789, the German Illuminati were thought by Pope Pius VI to be dedicated to the "total destruction of the Catholic religion and monarchies." One author denies that most French Masons supported the revolution. However, consider the near-bankruptcy of the French government around 1787 due to the wealthy "nobles" avoiding taxes. Involvement of some Freemasons was very complex (29)(30)(31).

The Jesuits (Society of Jesus), founded 1534-1540 AD/CE, "the pope's warriors," were supposedly not involved in this revolution, but were prominent in world-wide efforts to convert non-believers and "infiltrate Protestant groups (32)(33)."

Revolutions where Masons (e.g., Sir Thomas Moray) may have been involved, date at least from the English Civil War of 1640, and what would later be called the "bloodless revolution" of 1688 in England when Catholic King James fled to France from William of Orange; another French Revolution of 1830; and revolutions during 1848 in Sicily and Paris, in Berlin and in other parts of Germany, "to overthrow established religion and monarchies."

For hundreds of years many secret plots, power-seeking intrigues, and atrocities have involved Catholics versus Protestants and the pope versus monarchs. In 1612 AD/CE, English King James, a "secret Catholic," allowed his Protestant government to have Jesuits and other Catholic priests "hanged, drawn and quartered on trumped up charges."

In the late 1500s, the Jesuits, "soldiers of the Holy See," had been "very involved" in the assassination of the French king, Henry III. On and on, *ad infinitum*. Today, the intrigue continues (34)(35)(36).

American Masons generally consider their origins to have been in England and Scotland. In the early 1700s, Masonry quickly spread to Europe and the American colonies.

In 1786 AD/CE, a Masonic plan for world conquest by *any means*, to abolish Christianity and to assassinate monarchs has been described, as below.

In 1785 in New York City the Columbian Lodge of the Illuminati had been established with members including Governor DeWitt Clinton, Clinton Roosevelt and Horace Greeley (37).

The continued influence of Freemasonry, with its purported origins in ancient Egypt, can be seen today in its symbols, such as the "eye of the Osiris," on the back of the Great Seal of the United States and on the back of the U.S. one dollar bill. The City of Washington, D.C., and many government buildings and edifices reflect Masonic symbols, e.g., the Pentagon (38)(39)(40)(41).

Prominent Masons in history have purportedly included Francis Bacon (1561-1626), who wrote *Novum Organum* in 1620; Voltaire (1694-1778), who advocated tolerance in religion and politics; Jean Jacques Rousseau (1712-1778); King Frederick II of Prussia (1740-1786), who advocated religious tolerance; Wolfgang Mozart (1756-1791); Napoleon Bonaparte (1769-1821); Cecil Rhodes (1853-1902), who established the Rhodes scholar-ship; Winston

Churchill (1874-1965); Franklin D. Roosevelt (1882-1945) and Josef Stalin (1879-1953).

Many U.S. Presidents are said to have been Masons: George Washington, James Monroe, Andrew Jackson, James Polk, James Buchanan, Andrew Johnson, James Garfield, William McKinley, Theodore Roosevelt, William Taft, Warren Harding, Franklin D. Roosevelt, Harry Truman and Gerald Ford. Of the 37 Presidents before Jimmy Carter, either 18 or 21 were "close relatives."

President John Quincy Adams opposed Masonry as "childish, ridiculous, foolish, harmful and against the basic principles of democracy and equality and respect for the laws of the United States (42)(43)."

By 1900, the Freemasons in America had banned all political arguments in their lodges. Freemasons have been Republicans, Democrats, "liberals" and "conservatives." All major religious beliefs are tolerated. They salute the American flag during their meetings and are expected to abide by laws of their country.

Compare and contrast the following recent authors with Still, Marrs and Webster (44)(45).

Some present day "Knights Templar of Paris," an Order residing in the Jesuit College of Clermont, claim to be the true descendants of the ancient Knights, with the secret teachings of Egypt, Moses, the Israelites, and thence to the Knights Templar. They adhere strictly to the Bible and therefore, supposedly, cannot claim descent from the heretical Knights. "True Masons reject this spurious Order." The same source of this belief asserts the falsity of the virgin birth of Jesus and his misunderstood statements regarding being the Son of God. One very detailed account may be found elsewhere (46).

FREEMASONS AND THE KNIGHTS TEMPLAR

[REFS 47-78]

Freemasons have been linked to the Knights Templar or Poor Knights of the Temple of Solomon founded in 1118 AD/CE by the French knight, Hughes de Payens, to protect Christian pilgrims traveling through the Holy Land. These Knights Templar, like the Order of St. John and the Teutonic Knights, originated from the Christian Crusades against the Muslims in Jerusalem.

These secretive Knights of the Temple were so named for the ruins of Solomon's Temple. They "adopted the white habit of the Cistercians, adding to it a red cross." In the late 14th century, Christian Rosenkreutz allegedly founded the secret Order of the Holy Cross, the modern Rosicrucians, with ancient rites also incorporated into the Scottish Rite of Masonry (47).

In 1128, Pope Honorius II had recognized the Templars, with a "Rule" which provided the Order with a structure greatly resembling later Freemasonry. Thereafter the Templars rapidly attained great wealth and power throughout Europe, rivaling the Vatican. They were given extensive privileges by the Roman Church, including freedom from taxation by kings or the Church. (Incidentally, Honorius II became Pope in 1124, preceded by Gregory VIII in 1118 AD and Callistus II in 1119).

The Templars' sudden, rapid growth, as well as their "demise," beginning in France on the 13th of October in 1307, could be related to their alleged "secrets threatening the very foundations of the Church." This has been referred to as "hidden knowledge regarding the true story of Jesus": No virgin mother and no resurrection. There are other authors who would agree (48).

Note the parallels between the mythology of ancient Egypt: Isis (the holy virgin), mother of Horus, who was sired by the resurrected Osiris (god of the underworld). Also note that Judaism and Islam have never accepted the virgin birth nor the divinity of Jesus as the "son of God." (Julius Caesar was the first Roman emperor to be "deified" — by the Roman Senate in 42 BCE —and his nephew, Octavian, assumed the title "Son of God.")

This version of the Templars' secret "heresy" has been strenuously denied, even when the Templars' "confessions" were obtained by torture or such threats (49)(50)(51)(52).

The demise of the Templars seems related to the "genocide" of the heretical Cathars in southern France. Cathars denied the authority of the Pope. The Albigensian Crusade was declared by Innocent III in June 1209 AD/CE. The wholesale massacres continued until 1244. Many Templars lived in the Albi-Languedoc area and were also later considered to be heretics. In March 1244, rather than renounce their version of Christianity, some 205 Cathars were burned alive at fortified Montsegur in Southern France, after years of repelling efforts by the Church of Rome to convert them.

This represented a continuation of the 1209 "special crusade" by the Roman Church to this perceived tremendous threat. "Whole towns loyal to the Cathars were massacred in the most brutal fashion." In 1215, the Fourth Lateran Council "sounded a warning" and in 1231, Pope Gregory IX had established the Inquisition resulting in a million persons killed, more than in any other Crusade against "heretics."

This massacre of some 100,000 "heretical" Cathars in southern France, on the order of Pope Innocent IV, has been called "the first act of European genocide." (At Beziers in 1208 and at Albi in 1244 AD/CE.)

The earliest Christians are said to have considered John the Baptist the true "Messiah" and the "true succession," not Peter, as claimed by the Catholic Church. "Captured documents" indicated that the Knights Templar had always renounced "the religion of Peter" and had been secretly initiated into the Eastern Church or "Primitive Christian Church." John the Baptist was the patron saint of the Knights Templar and Freemasons.

The Knights Templar, Freemasons, the "heretical" Cathars, the Essene "dissenters" of the Dead Sea, and the cult of Isis and Osiris have all been interconnected by a number of writers (53)(54)(55).

To this date, alleged links between the Templars (1108-1307), the Syrian Assassins (circa 1200), the Cathars (1208-1244), Rosicrusians (c.1614), and Freemasons over a period of hundreds of years have been discussed in detail. Any link between Islam and Freemasonry and a New World Order remains obscure to this author at the present time.

In 1129, Baldwin II, the Christian King of the "Holy City of Jerusalem," so named after its capture in the first Crusade of 1096-1099, had asked the renowned Templar warriors to assist in the attack on Muslim Damascus. It failed, supposedly, due to a conspiracy between Templars and "and Islamic secret society," the Assassins. Their members inside Damascus are thought to have revealed the plot prior to the assault. Nevertheless, the "Knights of Jerusalem" were considered "the terror of the Mohammedans." In less than two centuries, some 20,000 Templars "perished in war (56)(57)."

By 1156 the Catholic clergy in the Holy Land sought relief from the "exorbitant privileges of the military orders," only to be set aside by the Holy See in Rome. Some sources consider the Templars to have preceded the German Rothschilds and the Medici of Florence in developing modern banking, handling of the credit and available capital in Western Europe. They could be seen as the "medieval equivalent of today's multi-national corporation(58)."

Many castles and cathedrals were built with their contributions including, in 1134, the magnificent Chartres Cathedral southwest of Paris. It was named after the Carnutes, a Celtic tribe, and erected on the site of an ancient Druid center. (See the historic novel, *Druids*, by Morgan Llywelyn, Ballantine Books, 1999.) Such grand works apparently involved the early stonemason guilds, with their knowledge of architecture, promoted by the Templars. Many cathedrals were built on pagan sites, e.g.; Notre Dame in Paris over the former temple of Diana and St. Sulpice over the ruins of a temple of Isis. Lengthy discussions are available regarding the similarity of these cathedrals to the structure of Solomon's temple and presence of symbols common to Masonry, ancient Greece and Egypt.

Nearing the year 1300, the Templars are credited with creating another military Order, "the formidable Teutonic Knights, the childhood heroes of Adolf Hitler." The Teutonic Knights' emblem was the swastika (59).

The extreme secrecy of the Teutonic Order gave rise to suspicions and allegations of "denial of Christ," "the worship of idols," and other possible "inventions (60)."

It is believed they were "crushed by an envious French king and a pope fearful of their secrets." In 1303, Philip IV had seized Pope Boniface VIII as part of a protracted power struggle between the "divine" monarchs and the "holy" Church (61) .

On Friday, October 13, 1307 AD/CE, the demise of the Templars by imprisonment, torture and burning at the stake began at the hands of the French Philip IV (not to be confused with Philip II of Spain, who massacred opponents of Catholicism during the Inquisition in the mid-1500s).Philip IV had finally convinced Pope Clement V that the Templars were plotting the overthrow of the Church. A painting depicts Jacques de Molay and Geoffrey de Charney being burned at the stake before the gates of Notre Dame in November 1314 (62)(63)(64).

In 1312, the Knights Templar were "dissolved" by Clement V, (the Pope *after* 1305), allegedly "as a mere formality," since their wealth and secrecy, continued, according to some writers. This decree was known as *Vox Clamantis* (War Cry), supposedly only to appease Philip IV, but actually "continuing to enrich the Papacy" through the financial dealings of the Templars.

There are conflicting opinions regarding the continuations of the Templars. Their property and some of their members were acquired by the Hospitaliers, the Knights of Rhodes and of Malta. There is reason to believe that some were able to survive, perhaps under new names. One fantastic theory has it that some escaped with their vanishing fleet and treasures, sailing from Portugal to America, or "la Merica," in 1308, "flying their well-known skull-and-crossbones battle flag (65)."

Another version: Prior to his execution, Grand Master, Jacques de Molay, sent 13 Templars to Stockholm, Naples, Paris, and Edinburgh to establish "new lodges." The Edinburgh lodge became the headquarters of Scottish Rite Freemasonry (66).

Again, a hidden reason for the sudden good fortune and later demise of the Templars allegedly was related to their possession from 1118 of secret evidence of the early Church having misrepresented Jesus' "virgin birth" and "resurrection" — among other things (67)(68).

In 1688, England's unpopular pro-Catholic king, James II, known as Jacobus in Latin, was deposed and fled to France. Freemasons in Scotland and Wales continued to support him in his claim to the English throne. With the help of the French king, Louis XI and the Catholic Jesuits, he established a system of Masonry called the "Scottish Rite." The Scottish Rite is said to be the most powerful Masonic order in the world (69)(70)(71)(72).

Hence, in Edinburgh, the Templars became the Scottish Rite Freemasons, who were later identified with the "American Freemasons." Many who signed the Declaration of Independence were from English Freemasonry.

In 1738, soon after additional evidence of the connection between the Knights Templar and Freemasonry, Pope Clement XII condemned Freemasonry as "pagan" and threatened with excommunication any Catholic who joined the Freemasons. Consider the case of John Coustos, an Englishman in Lisbon, in 1744, sentenced by the Inquisitors to torture and death for *illegally* establishing a Freemasons' lodge in Portugal. Although he "recanted," he was tortured, though not put to death, but sentenced to 5 years as a slave in the galleys.

More detailed descriptions are available regarding the close relationships between the Knights Templars and the Roman Church, Teutonic Knights, and the Scottish Rite of Freemasonry (73)(74)(75).

In Germany and Austria, some Templars became "Rosicrucians" and "Teutonic Knights." The latter are said to have formed the nucleus for Hitler's political support 600 years later. Adolf Hitler, Der Fuhrer, an alleged Catholic, in his later ravings, said he would hang the Pope if Germany won the war (WWII). This, despite a blessing, a Vatican Concordat, from the Pope, early in Hitler's political career. Hitler had called Christianity "an invention of sick brains." By 1942, persons close to him considered him to be "mad." Mein Kampf, written in 1925, allegedly with the aid of a Jesuit, later replaced the Bible on German Church altars (76).

In June 1943, as Hitler's SS troops were exterminating thousands of Jews, Pope Pius XII said he considered communism "a far greater danger" (to the Church) than Nazism (77)(78).

FREEMASONS AND THE VATICAN BANK

[REFS 79-80]

"The Vatican Bank has the distinction of being one of the most notorious and secretive financial institutions in the world." The bank was organized in 1941 by order of Pius XII, the so-called

"Hitler's pope." Allegedly, "Nazi Gold" and other funds were transferred from the Reichsbank to the *Instituto per le Opere di Religione* (IOR) and then to Nazi-controlled banks in Switzerland. The IOR has also been accused of involvement of the disappearance of Croatia's treasury of some $200 million in 1945 (79)(80).

In the late 1970s, the Bank allegedly was involved in Mafia money-laundering schemes, mostly orchestrated by American Bishop Paul Marcinkus. Many other bishops have been named, with some members of the "notorious P2 Masonic lodge." In one single transaction, some $95 million disappeared, "documented by the Irish Supreme Court."

FREEMASONRY AND ISLAM

[REFS 81-84]

On 15 July 1978, the El-Azhar University and the Islamic Jurisdictional College (IJC),issued an opinion, "after complete research," concerning Freemasonry (81). Their conclusions were:

1) Freemasonry is a clandestine organization, which conceals or reveals its system, depending upon circumstances. Its actual principles are hidden from members, except for chosen members of its higher degrees.
2) The members of the organization, worldwide, are drawn from men without preference for their religion, faith, or sect.
3) The organization attracts members on the basis of providing personal benefits. It traps members into being politically active, and its aims are unjust.
4) New members participate in ceremonies of different names and symbols, and are frightened from disobeying its regulations and orders.
5) Members are free to practice their religion, but only mem-

bers who are atheists are promoted to its higher degrees, based on how much they are willing to serve its dangerous principles.

6) It is a political organization. It has served all revolutions, military and political transformations. In all dangerous changes a relation to this organization appears either exposed or veiled.
7) It is a Jewish Organization in its roots. Its secret higher international administrative board includes Jews and it promotes Zionist activities.
8) Its primary objectives are the distraction of all religions and it distracts Muslims from Islam.
9) It tries to recruit influential financial, political, social, or scientific people to utilize them. It does not consider applicants it cannot utilize. It recruits kings, prime ministers, high government officials and similar individuals.
10) It has branches under different names as a camouflage, so people cannot trace its activities, especially if the name of "Freemasonry" has opposition. These hidden branches are known as Lions, Rotary and others. They have wicked principles that completely contradict the rules of Islam. There is a clear relationship between Freemasonry, Judaism, and International Zion. It has controlled the activities of high Arab officials in the Palestinian problem. It has limited its duties, obligations and Activities for the benefit of Judaism and International Zionism.

Finally, any Muslim who affiliates with it, knowing the truth of its objectives, is an infidel to Islam.

The same source of this information states that after Pope Clement XII in 1739 excommunicated Freemasons as atheists, the Sultan Mahmut also outlawed Freemasonry in the Ottoman

Empire. Freemasonry is outlawed in most Arab countries with the notable exception of Algeria, Lebanon and Morocco.

There are Grand Lodges in Morocco, Egypt and Turkey (81). Some 700 years ago (c. 1300 AD) two secret societies, the Knights Templar and the Syrian Assassins (Nizari Ismailis) were destroyed as heretics. Their interactions as "guardians of sacred mysteries" have been described in detail. "These holy warriors, ostensible opponents, are seen as mirror images." Secret and "occult" societies, ancient and modern, have been compared at great length.

Hasan bin Sadah, founder of the fanatical Assassins, an outgrowth of the Islamic sects of Hakim, Fatima, the Batinis and Shiahs, had earlier studied at the Dar ul Hikmat (House of Knowledge) or Grand Lodge in Cairo. There he learned "ancient knowledge," as in the Hebrew Cabala, later used in organizing the Assassins—and also used by Weishaupt and the Illuminati in Bavaria in 1776 (82)(83).

Abdullah ibn Maymun (c. 872 AD/CE) further prepared the way for development of the Assassins and as a Gnostic, planned to abolish all religions, including Islam, while posing as a "pious Ishmaili (84)."

FREEMASONRY AND JUDAISM

[REFS 85-95]

Around 100 BC/CE, a Jewish sect of some 4000, known as the Essenes, were in conflict with other Jewish sects, proposing that all property be communal, as in later "communism," and as with the earliest Christians. This "communal" living was proposed by Plato (c.427-347 BC/BCE) in *The Republic*. They were also considered to be the better educated classes of the Jews. Newer historical information regarding the Essenes has been found with the "Dead Sea Scrolls" or Qumran (85)(86)(87)(88)(89).

The Essene craftsman were thought by Manley Hall, a 33rd degree Mason, to be the "progenitors" of modern Freemasonry, e.g., from the builders of Solomon's Temple, hundreds of years earlier, dedicated to the new god, Yahweh. Solomon (961-922 BC/BCE), son of David, became King of Israel. David had captured Jerusalem from the Philistines and by 1000 BC/BCE established a kingdom (90).

The goat's head, as a pentagram, was a symbol of the Templars and seen in the Old Testament *cabala*. Yom Kipper, Day of Atonement, involved a goat sacrice. Note that the Knights Templar goat-idol, Baphomet, meant "absorption-into-wisdom."

Consider the U.S. military headquarters, the Pentagon, and the five-sided design of the city of Washington, D.C. (91)

Manley Hall, a prominent Mason and author, considered Gnosticism to be an integral part of the Ancient Mysteries—as in Masonry, Andre Nataf, an "occult author," is quoted as saying that, with Gnosticism .... "they could subscribe to the outward doctrines of any religion, and continue to operate under many different politico-religious systems (92)."

Gnosticism has been described as based on "secret knowledge" and the ability of human intelligence to determine truth from falsehood. *Cabala* is defined as a system of esoteric (secret or meant for a select few) theosophy (mystical religious philosophy) developed by medieval rabbis; any occult (dealing with magic; supernatural; available only to the initiated) doctrine (93).

Gnosticism "flourished as a religion until declared heresy by a council of bishops of the Roman Church in 325 AD/CE" at Nicaea (94)(95)

FREEMASONRY AND COMMUNISM

[REFS 96-106]

Masonry and Communism seem to have similar "utopian" ideals as did Plato. For many centuries, some kind of utopian world has been dreamed of. Plato (c. 427-347 BC/BCE) was one of the earliest philosophers to elaborate on this concept in his *Republic* and the *Laws*, written around 400 BCE (96)(97).

Around 10,000 BC/BCE, a utopian society of Atlantis supposedly conquered Greece and "all of Europe." This so enraged Zeus, the father of all gods, that he caused the "continent of Atlantis," with it 60 million inhabitants, to sink into the ocean in a single evening.

The *Atlantis* story may have been passed down orally from the Greek philosopher Solon, who had studied with Egyptian priests at the Temple of Isis around 595 BC/BCE. For Masons, America was to become the legendary "New Atlantis" of Plato (98)(99).

According to one prominent Masonic author, this has continued to be considered the "ideal pattern of government (100)."

The Bible "hinted at" the early Christians attempting to hold property in common, not as individuals (101).

"Communal living" was attempted by the early New England Puritans, with near disastrous results. William Bradford denounced the English merchants who financed the *Mayflower* for insisting on this "communal living" of Plato's (102).

The most horrendous example of *forced* communism was surely in the Soviet Union under Stalin (103).

The French Revolution, has been considered as "liberal" rather than "socialist," but like many revolutions throughout history, it represented a revolt of the poor and subjugated against the wealthy ruling "masters." Revolutionaries in France included the "more extreme Jacobins," in contrast to most Masons (104)(105).

The origin of the term "Jacobins" is believed to refer to a secret organization in France. When the unpopular Catholic king of England, James II, was deposed and fled to France in 1688 AD/CE, Freemasons in Scotland and Wales tried to return him to the English throne. French Freemasons accused them of converting Masonic rituals into a political cause. (James, in Latin, was Jacobus (106).)

REFERENCES AND NOTATIONS: CHAPTER SEVEN

THE CANDIDATES

1) Marrs, Jim. *Rule of Secrecy.* New York. HarperCollins Publishers, 2000.[Part One: Modern Secret Societies]
2) Ross, Robert Gaylon, Sr. *Who's Who of the Elite: Members of the Bilderbergs, Council on Foreign Relations & Trilateral Commission.* Spicewood, TX.RIE Publishers, 2000.
3) McManus, John. *The Insiders.* Appleton, WI. The John Birch Society, 1992.
4) Webster,Nesta.*Secret Societies and Subversive Movements.* Hawthorne, California. Christian Book Club of America. First Published 1924.
5) MacKenzie,Norman (Editor).*Secret Societies.* London. Aldus Books, 1967.
6) Marrs,Jim. *Rule by Secrecy.* New York. HarperCollins, 2000.
7) Still, William. *New World Order: The Ancient Plan of Secret Societies.* Lafayette, LA. Huntington House, 1990. 141

SKULL AND BONES

8) Sutton,Antony. *America's Secret Establishment: Introduction to the Order of Skull and Bones.* Billings, Montana. LibertyPress, 1986.
9) Robbins,Alexandra. *Secrets of the Tomb: Skull and Bones, The Ivy League and the Hidden Paths of Power.* Boston, New York, London. Little, Brown, 2000.
10) Still, William. *New World Order.* Lafayette, LA. Huntington House Publishers, 1990. [Pages 39 and 82]
11) Mackenzie,Norman (Editor).*Secret Societies.* London. Aldus Books Limited, 1967.[See pages 129 and 307]
12) Picknett,Lynn;Prince,Clive. *The Templar Revelation: Secret Guardians of the Identity of Christ.* New York. Touchstone/Simon & Schuster,1997.
13) Sutton,Antony. *America's Secret Establishment: An Intro- duction to the Order of Skull and Bones.* Billings, Montana. Liberty Press 1986.
14) Still,William. *New World Order.* Lafayette, LA. Huntington House Publishers,1990.[See page 73]
15) Goldstein,Paul and Steinburg,Jeffrey. *George Bush, Skull and Bones and the New World Order.* A New American View, International Edition, April 1991. [See wwwparascope.com/articles/0997/white paper.htm]
16) Robbins, Alexandra. Secrets of the Tomb. Skull and Bones, the Ivy League, and the Hidden Paths of Power. Boston, New York. London. Little, Brown, and Company, 2000. [Also see "Secret Origins of Skull

and Bones," viewed at http://www.parascope.com AND "Skull and Bones," viewed at www.geocities.com/CapitolHill]

17) Marrs, Jim. *Rule of Secrecy: the Hidden History that Connects the Trilateral Commission, the Freemasons and the Great Pyramids.* New York. HarperCollins Publishers, 2000. [Quoted on page 13]

ILLUMINATI

18) Still,William. *New World Order.*[See pages 37-39]
19) Saussy,F. Tupper. *Rulers of Evil.* New York. HarperCollins, 1999.[See pages 296-297 for Jesuit Superior Generals]
20) Garraty and Gay. *Columbia History of the World,* 1972. New York, London. Harper & Row,1972. [See page 543]
21) Saussy,F. Tupper. *Rulers of Evil,* 1999. [pages 40 and 171]
22) MacKenzie, Norman (Editor). *Secret Societies.* London. Aldus Books, 1967.[See pages 169 and 300]
23) Still,William. *New World Order,*1990. [Chapter Five: Weishaupt's Illuminati]

THE FREEMASONS

24) Leadbetter,Charles W. *Freemasonry and Its Ancient Mystic Rites.*New York.Gramercy Books,1986.Original publication, 1926.[See pages 163-168]
25) Ridley,Jasper. *The Freemasons:A History of the World's Most Powerful Secret Society.* New York. Arcade Publishing, 1999,2001.
26) See www.grandlodgeoftexas.org/165years.html
27) Leadbetter,Charles W. *Freemasonry and Its Ancient Mystic Rites.* New York. Gramercy Books,1986.Original publication, 1926.[See pages 163-168.This is an extraordinarily detailed history of Masonry]
28) Still, William. *New World Order.* Lafayette, LA. Huntington House Publishers, 1990.[Pages 50 and 52]
29) Ridley, Jasper. *The Freemasons: A History of the World's Most Powerful Secret Society.* New York. Arcade Publishing, 2001.[See Chapter 11]
30) Garraty and Gay. *Columbia History of the World,* 1972. New York, London. Harper & Row, 1972 See chapter 88*: the United States in Prosperity and Depression.*
31) Still,William. *New World Order.* Lafayette, LA. Huntington House Publishers, 1990.
32) Paris, Edmund. *The Secret History of the Jesuits.* Ontario, CA. Chick Publications, 1975.
33) Columbia History of the World, 1972. [Page 334 and 712]

34) Paris,Edmund. *The Secret History of the Jesuits.* 1975. Ontario, CA. Chick Publications.[See Section II, Chapter 6]

35) Marrs,Jim. *Traitors, Treason and Treachery: The French Revolution* available at www.freemasonwatch.freepress-freespeech.com

36) Marrs,Jim. *Rule of Secrecy: the Hidden History that Connects the Trilateral Commission, the Freemasons and the Great Pyramids.* New York. HarperCollins Publishers, 2000. 143

37) Webster,Nesta. *World Revolution.* Devon. United Kingdom. Britons Publishing Company, 1921 and 1971.

38) Saussy,F.Tupper.*Rulers of Evil.* New York. HarperCollins, 1999.[See Chapter 20: *American Graffiti*]

39) Still,William. *New World Order, 1990.*[Chapter 8: American Masonry]

40) Marrs, Jim. *Traitors, Treason and Treachery: The French Revolution* available at www.freemasonwatch.freepress-freespeech.com]

41) Howard, Robert. *United States Presidents and the Masonic Power Structure*, 1971. Original publication 1921.

42) Bessel, Paul M. *Masonic and Anti-Masonic Presidents of the United States.* Available on the Internet at www.bessel.org/presmas.htm]

43) Howard, Robert. *United States Presidents and the Masonic Power Structure*, 1971. Original publication 1921

44) Ridley, Jasper. *The Freemasons: A History of the World's Most Powerful Secret Society.* New York. Arcade Publishing, 2001.

45) Lomas, Robert. *Freemasonry and the Birth of Science.* New York. Barnes and Noble, 2002.

46) Theosophy, 1970,found at www.wisdomworld.org/additional/christianity/JesuitryAndMasonry4of6.html]

FREEMASONS AND KNIGHTS TEMPLAR

47) Ridley,Jasper. *The Freemasons: A History of the World's Most Powerful Secret Society.* New York. Arcade Publishin, 1993,2001.[See pages 13-14,16,21;22]

48) Reuchlin, Abelard. *The True Authorship of the New Testament: Flavius Josephus.* Kent, WA. Abelard Foundation,1979.

49) Ridley, Jasper. *The Freemasons.* New York. Arcade Publishing, 2002.

50) Picknett,Lynn;Prince,Clive. *The Templar Revelation: Secret Guardians of the True Identity of Christ.* New York. Touchstone, 1997.[See Chapter Four]

51) Bauval, Robert;Gilbert,Adrian. *The Orion Mystery.* New York. Three Rivers Press, 1994.

52) Garraty and Gay. *Columbia History of the World.* New York, London. Harper & Row,1972.[See page 91 and page 216]

53) Westerman,James. *The Templars and the Assassins: the Militia of Heaven.* Rochester, Vermont. Inner Traditions, 2001. [See Chapter 9,19,24,25,and on pages 194, 263, 270- 272]

54) Caroll,James. *Constantine's Sword: The Church and the Jews.* Boston. Houghton Mifflin, 2001.[See pages 306,636]

55) Peters,F.E. *Children of Abraham.* New Jersey. Princeton University Press, 1982.

56) Moeller, Charles. *The Catholic Encyclopedia: The Knights Templar.*[Found at www.newadvent.org/cathen/14493a.htm]

57) Marrs, Jim. *The Templar Revelation: Secret Guardians of the True Identity of Christ.* New York: Touchstone Books, 1997.[See page 278]

58) Marrs,Jim. *The Templar Revelation,*1997.[See page 285]

59) Marrs,Jim.*The Templar Revelation.*[Pages 159 and 304]

60) Moeller Charles. *The Catholic Encyclopedia: The Knights Templar.*[www.newadvent.org/cathen/14493a.htm]

61) Marrs,Jim. *The Templar Revelation,* 2000.[Previously Cited at Ref 58. See page 274]

62) MacKenzie, Norman (Editor). *Secret Societies.* London. Aldus Books, 1967.[Page 128]

63) Saussy,F. Tupper. *Rulers of Evil* New York .HarperCollins, 1999.[Pages 37-40]

64) Leadbetter, Charles W. *Freemasonry and Its Ancient Mystic Rites.* New York.Gramercy Books,1986.Originaly Published, 1926.[Pages 163-168]

65) Marrs,Jim. *Rule by Secrecy,*2000. [Page 307]

66) Saussy,F. Tupper. *Rulers of Evil* New York. HarperCollins, 1999. [Page 40]

67) Picknett, Lynn, Prince, Clive. *The Templar Revelation: Secret Guardians of the True Identity of Christ.* New York. Touchstone, 1997.[Chapter Eleven: Gospel Untruths]

68) Marrs,Jim. *Rule by Secrecy: the Hidden History that Connects the Trilateral Commission, the Freemasons and the Great Pyramids.* New York. HarperCollins Publishers, 2000.[See page 305]

69) Still, William. *New World Order: The Ancient Plan of Secret Societies.* Lafayette, LA. Huntington House Publishers, 1990. [Page 109]

70) Marrs, Jim. *Rule by Secrecy: the Hidden History that Connects the Trilateral Commission, the Freemasons and the Great Pyramids.* New York. HarperCollins Publishers, 2000. [See page 223]

71) Leadbetter,Charles W. *Freemasonry and Its Ancient Mystic Rites.* New York. Gramercy Books 1986.[Chapter Eleven: The Scottish Rite]
72) Ridley,Jasper. *The Freemasons: A History of the World's Most Powerful Secret Society.* New York. Arcade Publishing, 2001.
73) Leadbetter,Charles W. *Freemasonry and Its Ancient Mystic Rites.* New York. Gramercy Books, 1986. Original publication, 1926. [Pages 172-177]
74) Picknett, Lynn, Prince, Clive. *The Templar Revelation: Secret Guardians of the True Identity of Christ.* New York. Touchstone, 1997.
75) Ridley, Jasper. *The Freemasons: A History of the World's Most Powerful Secret Society.* New York. Arcade Publishing, 2001. [Pages 55-56]
76) Bromberg,Norbert and Small,Verna. *Hitler's Psychopath- ology.* New York. International Press, 1983.[See page 147]
77) Toland, John. *Adolf Hitler.* New York. Ballantine Books, 1976. [Page 549]
78) Paris,Edmund. *The Secret History of the Jesuits.* 1975. Ontario, CA. Chick Publications. [See Chapter 7]

FREEMASONS AND THE VATICAN BANK

79) Levy, Jonathan. *The Vatican Bank.* In Kick, Russ (Editor). *Everything You Know is Wrong: The Disinformation Guide Secrets and Lies.* New York. The Disinformation Company, Ltd., 2002.
80) Saussy,F.Tupper. *Rulers of Evil* New York. HarperCollins, 1999.[*The Vatican Treasury,* page 160]

FREEMASONS AND ISLAM

81) Layiktez,Cecil. *Freemasonry and the Islamic World.* Tesvyye, *Masonic Magazine of the Grand Lodge of Turkey.* [www.users-libero.it/fjit.bvg/layiktezl.html] [ALSO see Mustafa, El-Amin. *Freemasonry, Ancient Egypt and the Islamic Destiny*]
82) Marrs,Jim. *Rule of Secrecy: The Hidden History that Connects the Trilateral Commission, the Freemasons and the Great Pyramids.* New York.HarperCollins,2000.[P.235}
83) Wasserman,James. *The Templars and the Assassins.* Rochester, VT. Inner Traditions International, 2001. [See pages 11,183,199,223,263]
84) Marrs,Jim. *Rule of Secrecy: The Hidden History that Connects the Trilateral Commission, the Freemasons and the Great Pyramids.* New York.HarperCollins,2000.[See page 281]

FREEMASONS AND JUDAISM

85) Shanks,Hershel.*The Mystery and Meaning of the Dead Sea Scrolls.* New York. Random House, 1998.

86) Schiffman,L. *The Halakhah at Qumran.* Leiden. E.J. Brill, 1975.
87) Peters, F.E. *Children of Abraham: Judaism-Christianity- Islam.* Princeton University Press, 1982.[See pages 13-14]
88) Armstrong,Karen. *History of God,* 1993 [See pages 71-80]
89) Garraty and Gay. *Columbia History of the World.* New York & London. Harper & Row, 1972.[See page 217]
90) Garraty and Gay. *Columbia History of the World.* New York & London. Harper & Row, 1972,[See pages 141-142]
91) Saussy,F.Tupper. *Rulers of Evil.* New York. HarperCollins, 1999.
92) Marrs,Jim. *Rule of Secrecy.* New York. HarperCollins, 2000.
93) *The Random House Dictionary.* New York, 1980.
94) Saussy,F.Tupper. *Rulers of Evil.* New York. HarperCollins, 1999.[See Chapter Four: *Medici Learning* and pages 23-24]
95) Leadbetter,Charles. *Freemasonry and It's Ancient Mystic Rites.* New York. Gramercy Books, 1986. Original Publication 1926.[See pages 163-168]

FREEMASONRY AND COMMUNISM

96) Armstong, Karen. *A History of God, 1993.* [See pages 34, 175, 201, 204, 312]
97) Durant, Will. *The Story of Philosophy.* New York. Simon and Schuster, 1926-1961. [Chapter One: Plato]
98) Hall, Manly. *America's Assignment with Destiny.* Los Angeles. Philosophical Research Society, 1973.
99) Still,William. *New World Order.* Lafayette, LA. Huntington House Publishers, 1990. [Chapter Three]
100) Hall,Manly. *Secret Destiny of America.* Los Angeles. Philosophical Research Society, 1944 and 1978.
101) Garraty and Gay. *Columbia History of the World,* 1972. New York, London. Harper & Row, 1972.[See pages 884-893]
102) Still,William . *New World Order.* Lafayette, Louisiana. Huntington House Publishers,1990.[See Early America, pages 56-59]
103) Garraty and Gay.*Columbia History of the World,* 1972. New York, London. Harper & Row, 1972.[See Chapter 87: Russian Revolution and Stalin Era]
104) Ridley,Jasper. *The Freemasons: A History of the World's Most Powerful Secret Society.* New York. Arcade Publishing,2001. [See Chapter 11]
105) Still, William. *New World Order.* Lafayette, LA. Huntington House Publishers, 1990.[Chapter Six – French Revolution; Chapter Seven- American Jacobins, Pages 83- 97]

106) Marrs, Jim. *Rule by Secrecy: the Hidden History that Connects the Trilateral Commission, the Freemasons and the Great Pyramids*. New York. HarperCollins Publishers, 2000.[Page 223]

SHADOW GOVERNMENTS

THE SHADOW GOVERNMENTS

[REFS 1-7]

Shadow Governments are global organizations, legal or illegal, which have accumulated immense wealth and political influence. They include the Federal Reserve System, inter-national banks, multinational corporations, the Bilberberg Group, Trilateral Commission, Council on Foreign Relations, arms dealers, drug cartels, international criminal enterprises and, probably, some powerful religions.

Numerous authors have described the *deliberate instigation of wars and economic crises* by international bankers and multi-national corporations, especially the arms makers and money lenders. Only they profit by every war. Historically, they have assisted both sides. The bloodshed is that of the deceived innocents.

The *tangled web of secrecy and deception* should become evident from the references provided next for the *Shadow Governments* (1)(2)(3)(4)(5)(6)(7) .

All wars involve *crimes-against-humanity.* The perpetrators of *invasions* often ignore the historical *Laws-of-war* (5)(6)(7) .

Every war involves tremendous budget deficits by all combatants. The borrowed money must be repaid by the middle-class taxpayer. Historically, many wars of conquest or *empirebuilding* have been conducted in the guise of some *true-religion,* while their true purpose was obtaining power over others, coupled with pure greed. Looting, burning, rape, torture, mass murder and the taking of slaves have been the standard for thousands of years.

In the U.S., the Federal Reserve System (a private corporation which controls the U.S. dollar), must be suspect. Other individuals and organizations are worthy of in-depth study and include the *military-industrial-complex* (so named by Dwight Eisenhower and implied by Abraham Lincoln), the *Committee of 300*,the multimedia giants, the super-wealthy Rockefellers, Morgans, Carnegies, and of course, the Rothschild international bankers.

The following detailed information concerns the Federal Reserve System, international banking, multinational corporations, the Bilderberg Group, the Trilateral Commission and Council on Foreign Relations.

FEDERAL RESERVE SYSTEM

[REFS 8-13]

This is actually a private bank, owned by banks, and as a "money monopoly" was considered by Thomas Jefferson (U.S. President (1801-1809) to be "the greatest danger to the survival of the Republic." Together with James Madison and Andrew Jackson, Jefferson argued that "the Republic and the Constitution were always in danger from the so-called 'money power', a group of autocrats, an elite we would call them today, who have manipulated the political power of the state to gain a monopoly over the money issue (8)."

On the other hand, Alexander Hamilton believed in a strong central government and a central bank overseen by a wealthy elite, based on the Bank of England, which was described as a "partnership between the government and banking interests."

At one point Hamilton was referred to as "a tool of the international bankers."

Thomas Jefferson opposed Alexander Hamilton's proposals for "a central bank, modeled after the bank of England, and overseen by a few wealthy elites," believing it to be unconstitutional and "more dangerous than a standing army."

After a Bill was introduced into the U.S. House of Representatives in December 1790, the First Bank of the United States was incorporated. This was the first privately owned banking monopoly in the U.S. Its capital was furnished 20 percent by the federal government and 80 percent by the Rothschilds and other foreign bankers. This Bank of the U.S. had a sole right to issue currency and "was exempt from taxation. The bank then caused inflation by creating fractional-reserve notes." The 'Money merchants prospered but refused to renew its 20-year charter.

John Adams is quoted as saying in 1811, "Every bank of discount, every bank by which interest is to be paid or profit of any kind made by the lender, is downright corruption. It is taxation of the public for the benefit and profit of individuals (9)(10)."

In 1816, the high cost of the War of 1912 caused Congress to issue a new charter to the *Second Bank of the United States.*

In 1832, President Andrew Jackson vetoed a bill to extend its charter, beginning the "Bank War." Jackson is quoted as saying that a central bank, in addition to being unconstitutional, was "a curse to a republic; inasmuch as it is calculated to raise around the administration a money aristocracy dangerous to the liberties of the country."

In 1835, America's first assassination attempt was made on President Andrew Jackson by Richard Lawrence, who claimed that he was "in touch with the powers in Europe." He was rumored to have been an agent of Jacob Rothschild in Paris.

After the assassination attempt, Jackson was so infuriated that he withdrew government funds from the "den of vipers." The Second Bank president, Nicholas Biddle, retaliated by curtailing credit nationally, causing widespread economic panic. The Bank's charter was not renewed.

Abraham Lincoln is considered to have been the last president to oppose the private banking system, but the Union government was hard-pressed to finance the Civil War. He had Congress print paper money(notes)to pay government debts without allowing private money to profit.

Bankers drafted the National Bank Act of 1863, which gave bankers control over the printing of money (11).

Several attempts to renew a central bank failed until 1913, when the Federal Reserve System was created. This had been secretly planned in 1910 on J.P. Morgan's Jekyll Island. The names of the schemers have been listed elsewhere. The "tangled web" of this conspiracy is described at length in a number of publications (12).

In 1913, Congressman Lindberg said that the Federal Reserve System "establishes the most gigantic trust on earth ....the new law will create inflation whenever the trusts want inflation. From now on depressions will be scientifically created."

Today, the Federal Reserve System consists of twelve Federal Reserve Banks, dominated by the New York Federal Reserve Bank. The majority of shares are owned by two banks, the Chase Manhattan Bank and Citibank.

"In the U.S. the control of money is by bankers of the Federal Reserve System. Money fuels the modern world (13)."

A further detailed history of the maneuverings of the American central bank is available. Note that the inter-national banks, e.g., the House of Rothschild, assisted Napoleon and his British enemies when they were at war with each other. Who really profited? This is just the up of the iceberg.

INTERNATIONAL BANKING

[REFS 14-18]

Between 1800-1850, "atypical" Jewish investment bankers prospered and proliferated, especially in Europe. They included the families of Oppenheim, Seligman, Warburg, Sasoon, and especially, the Rothschilds (14)(15).

Reportedly, Max Warburg, manager of a Rothschild-related bank in Frankfurt and a member of the German secret police in WW I, was the brother of Paul Warburg, a founder of the U.S. Federal Reserve System.

What are the sources for the enormous funding available for global conquest? Global arms sales and drug trafficking and other international criminal enterprises are tremendously profitable to someone.

Consider the demise of public property in South American countries following intolerable loan "conditions" by the World Bank (16).

Some evidence points to the ongoing methodical looting of developing nations through loans by the IMF and World Bank. There is a very different version provided by David Rockefeller in his recent "Memoirs."

David Rockefeller's recent autobiography reads like an eyewitness to momentous historical events in the twentieth century. He provides a detailed and credible description of the origins and actions of the Council on Foreign Relations, Trilateral Commission, the Bilderberg group, the World Trade Organization, International Monetary Fund, World Bank, and the North American Free Trade Agreement (NAFTA). No mention of Yale's "Skull and Bones" fraternity nor the Freemasons is made in his *Memoirs.* Despite other evidence, there seems very little doubt concerning his devotion to the United Nations and to a peaceful and healthy future for mankind. Perhaps only his choice of methods to this end can be questioned (17).

Compare Jim Marrs detailed and different appraisal of the Rockefeller family (18).

MULTINATIONAL CORPORATIONS

[REFS 19-26]

Global corporations have been acting in the name of improving the standard of living, the elimination of poverty, disease and ignorance around the world. Results have been much disputed.

Corporate Accountability International (CAI) in Boston, lists at least six "massive" international corporations who bankroll politicians who, in turn, "look the other way", while the corporations seriously endanger world citizens and the planet. As a result of past efforts of CAI, a number of these corporations have made reforms and been removed from the "Corporate Hall of Shame (19) ".

Over a period of multiple generations, both the Rockefeller and Rothschild families have had an enormous impact on world history, mainly through banking and the acquisition of enormous corporate assets. They have repeatedly been accused of starting and stopping wars, revolutions, economic crises, and depressions—all these allegedly, through the ability to control the world money market.

The relationships between multinational corporations, the World Bank and the International Monetary Fund are unclear, but grave accusations have been made regarding their alleged utopian goals versus to have been a systematic "looting" of some less developed countries.

One global process involves loans (with disastrous "conditions") by the World Bank and/or the International Monetary Fund (IMF) to countries already on the brink of economic disaster. (It may be significant that China has turned down offers of loans by the IMF. Could China replace the U.S. as "the richest nation on Earth"? Review the next references (20)(21)(22)(23).

A sign of the times may be evident in an article (October 6,2002) by Patrice M. Jones of the Chicago Tribune. In Brazil a leading candidate for the presidency was Luis Ignacio "Lula" da Silva. The odds were that he would win against President Fernando Henrique Cardoso due to the extreme anger the citizens felt over the "unfulfilled promises of a decade of "free-market reforms." Brazil was racked with rising debt, rampant crime, continued high unemployment. Much was blamed by da Silva on the "servile" relationship of South American countries on the United States and the Bush administration's interventionist policies. He saw that the Free Trade Areas of the Americas (FTAA), "as ... proposed by the United States," was not an integration proposal (but) actually "annexation politics."

Especially worthy of study is the David Rockefeller Center for Latin American Studies – Harvard University. Its stated purposes are very laudable. It "promotes the study of Latin America in all its dimensions (24)."

The Anglo-American Invasion of Iraq, begun in March 2003, was protested by millions of citizens in countries around the world, and opposed by the U.N. Security Council, by France, Germany, Russia and China, and others. The United States Congressional Budget Office (CBO) predicted that President Bush's 2004 budget will produce deficits over the next ten years of $1.8 trillion.

In early 2003, the U.S. national debt was $6.4 trillion. Congress was asked by the Treasury Department, for the second time in nine months, to raise the legal debt limit. At the same time, the U.S. government revenues continued to drop "dramatically", from

$2.02 trillion in 2000 to $1.99 trillion in 2001 and to $1.85 trillion in 2002. This was surely related to corporate mergers, massive corruption and fraud by big corporations and Wall Street brokers. Many brokerage firms continued to recommend "bankrupt" companies. Tax cuts for the wealthy were "insult added to injury." Most of U.S. States were also suddenly on the verge of bankruptcy and the President told them they were "on their own (25)."

Since the invasion of Iraq, I have heard no discussions regarding the question of who would pay for this "war." All wars are fought with borrowed money. What banks are to be repaid by the U.S. government, by way of those who actually pay taxes? An apparently deliberate devaluation of the U.S. dollar may be necessary to help reduce massive U.S. debt (26).

What motivates the United States in "liberating" third-world countries?

1) Are the Afghanistan and Iraq wars mostly about the continuing personal greed of multi-billionaires as the "globalizing" corporate owners?
2) Is the motivation, the "ancient plan" for a peaceful, prosperous and healthy world, known as the "New World Order?

In the opinion of persons who should know, *Globalization* has, so far, apparently caused more harm than good for the *developing* or poor countries, destroying their environments and exhausting natural resources.

A "New World Order," if based on the principles of the U.S. Constitution and the United Nations Charter, can surely never be realized by *forcing* these principles upon countries currently dominated by *tyrants,* whether in the form of religious fanatics, greedy "businessmen" or corrupt politicians.

THE BILDERBERG GROUP

[REFS 27-30]

This secretive group, "ruling the world since 1954," according to one source, meets once a year in May or June. It was created by Joseph Retinger (a secretary to novelist Joseph Conrad), David Rockefeller, Denis Healy, and Prince Bernard of the Netherlands. The organization was named for their first meeting in the Bilderberg Hotel in Oosterbeek, Holland, in 1954. The first meeting in the U.S. was in February 1957 on St. Simons Island near Jekyll Island, Georgia. Meetings rotate between countries. Each country has two steering committee members. Although "highly regarded American media members" meet with the Bilderbergers, little or nothing gets reported, as required.

Persons attending these meetings have included Prince Charles; David Rockefeller (chairman of Chase Manhattan Bank, net worth $2.5 billion); Umberto Agnelli of Italy's Fiat (net worth $3.3 billion); Vernon Jordon (credited with getting James Wolfensohn the presidency of the World Bank); James Wolfensohn; Henry Kissinger; Richard Holbrooke (former U.S. representative to the U.N.); Conrad Black (media magnate); Peter Mandelson (one time aide to Tony Blair); Lord Denis Healy (prominent British legislator and a Bilderberg founding member); Margaret Thatcher; U.S. presidents Regan, Bush (senior), Ford, Carter, and Clinton; Richard Cheney; Alan Greenspan (Federal Reserve); Caspar Weinberger; George Shultz; Walter Cronkite and Clint Eastwood. Many others, unnamed, have included "CEOs of pharmaceutical giants, tobacco companies, car manufacturers, international banks (27).

Jon Ronson, "Jewish journalist," quotes Denis Healey in describing Bilderberg members: "To say we were striving for a one-world government is exaggerated but not wholly unfair. Those of us in Bilderberg felt we couldn't go on forever fighting one another for nothing and killing people and rendering millions homeless.

So we felt that a single community throughout the world would be a good thing." Also, "Bilderberg is a way of bringing together politicians, industrialists, financiers, and journalists. Politics should involve people who aren't politicians. We make a point of getting along younger politicians who are obviously rising, to bring them together with financiers and industrialists who offer them wise words. It increases the chance of having a sensible global policy."

Jon Ronson also described in detail a 4-day conference, in July 2000, which he "infiltrated" at the Bohemian Grove (by the Russian River, near Occidental and Santa Rosa, California.

It included an evening at the lagoon with a symphony orchestra, speeches, drinking, dancing, urinating on the redwood trees, and naked swimming. Some attendees were in "drag." An eerie pagan-like ceremony by "priests in robes" in front of a 50 foot stone owl was led by the "High Priest," Jay Jacobus. The "voice of the owl" was John MacAllister (28).

An Investigative reporter, James P. Tucker, who followed the Bilderbergers for years, wrote: "The Bilderberger agenda is much the same as that of its brother group, the Trilateral Commission. The two groups have an interlocking leadership in a common vision of the world."

One author names "the three most notorious...modern secret societies" aas: the Trilateral Commission, the Council of Foreign Relations, and the Bilderbergers. David Rockefeller founded the Trilaterals but shares power in the older Bilder-berger group with the Rothschilds of Britain and Europe (29).

Recall that David Rockefeller has a very different view of the alleged "three most notorious and secret societies (30)."

THE TRILATERAL COMMISSION

[REFS 31-33]

This non-governmental, very private, Trilateral Commission

was organized in July 1972 and officially founded in July 1973 with David Rockefeller as chairman. It had been preceded by the more secretive Council on Foreign Relations (CFR), but there was an apparent need to be "more public."

The Commission was initiated by David Rockefeller and based on ideas of Zbigniew Brzezinski (head of Columbia University Russian Studies). In Brzezinski's writing he had seen "Marxism as a victory of reason over belief as the plan." In the 1970 CFR *Foreign Affairs* publication he is quoted as saying "A new and broader approach is needed—creation of a community of the *developed* nations which can effectively address itself to the larger concerns of man kind... A council representing the United States, Western Europe, and Japan, with regular meetings of the heads of governments as well as some small standing machinery, would be a good start. Headquarters are in New York, Paris and Tokyo and membership remains at about 300. A list of members is available but they are not to discuss the proceedings (31)."

The goals of the Trilateral Commission have been stated as "closer cooperation among the three regions on common problems." Some observers have seen it as "collusion" of multinational bankers and the corporate elite...toward one-world government.

Funding, according to a 1978 report covering mid-1976 to mid-1979 included $1.2 million from the tax-exempt Rockefeller Brothers Fund, plus contributions from the Ford Foundation, Lilly Endowment, *German* Marshall Fund, Time, Bechtel, General Motors, Wells-Fargo, Texas Instruments. More detailed and lengthy descriptions of this organization are available (32)(33).

COUNCIL ON FOREIGN RELATIONS

[REF 34]

The idea of a one-world community in the post-WW I era was discussed in New York in 1917 by "about 100 prominent men," including Col. Edward Mandell House, adviser to President Woodrow Wilson. Their plans "evolved" into the President's "fourteen points" for a peace settlement, presented to Congress in January 1918. They called for the removal of "all economic barriers" between nations, "equality of trade relations" and the formation of "a general association of nations." In 1912, Colonel House had talked of a "conspiracy" within the United States to establish a central bank, a graduated income tax and control of *both* political parties (34).

The Treaty of Versailles in 1919 resulted in such harsh demands for war reparations from Germany that, many believe, they led directly to the rise of Nazism. A League of Nations was proposed by President Wilson but rejected by the U. S. Senate.

Nevertheless, in the aftermath of the rejection, in July 1921, in New York, a branch of the British *Royal Institute of International Affairs* was organized and called the *Council on Foreign Relations.* Their by-laws provided for loss of membership for anyone discussing details of the meetings. Membership in the "by-invitation only" club is said to include 3,300 members today. A number of prominent members are said to hold many powerful positions in government (including the CIA), banking and industry. CFR members played a "key role" in American policy during and since WW II and have promoted the concept of a "global community."

MERCHANTS OF DEATH

[REFS 35-45]

ARMS DEALERS

The United Nations 56th and 57th General Assembly agendas (2001-2003) continued the debates over world arms sales, including land mines and small arms as well as nuclear disarmament, missile defense, weapons in space, biological and chemical weapons.

Most disturbing are the global statistics for arms sales. Between 1996 and 1999, the United States had continued to lead all others in international arms sales. The U.S. made $25.7 billion worth of arms sales agreements, Russia $14.3 billion and France $9 billion. Between 1992 and 1999, the largest recipients of arms sales were Saudi Arabia, Taiwan, the United Arab Emirates, China, Egypt, Israel, Kuwait, Malaysia and Pakistan. Russia's main customers were China and India, both at odds with Islam.

Using "a different numerical base," between 1995-1999, the top four suppliers were the U.S. at $53.4 billion, Russia with $14.6 billion, France at $11.6 billion and the United Kingdom at $7.3 billion (35).

In 2006, the United States continued to be the leading supplier of weapons to the *developing world,* followed by Russia and Britain. The top buyers were Pakistan, India and Saudi Arabia.[Source: New York Times News Service:10/01/2007]

After the first Persian Gulf War (August 1990-February 199 ended, the U.S. economy was "in the doldrums" and *cutting arms sales did not please the arms industry or* "labor."

Following the defeat of President George Bush, Senior, U.S. arms sales more than doubled during Bill Clinton's first year in office (1993-1994) (36)(37)(38)(39).

DRUG CARTELS

COCAINE: The more than 35-year old Colombian "civil war" has been a continuing item on the United Nations agenda as a "peace-making" effort. The most obvious underlying issue is poorly addressed. These conflicts are fueled by cocaine. Monstrous "crimes against humanity" continue. The most lucrative customer is the United States, where a so-called "war against drugs" has so far looked like a farce. To understand seemingly idiotic and harmful behaviors, the old rule applies-*follow the money.*" The war-on-drugs has received more attention recently with the discovery of vast oil reserves in territory controlled by rebels and supported by drug lords.

It should be noted that an explosive increase in cocaine refining in Columbia followed the construction of the Pan- American Highway ("cocaine highway") funded by the World Bank. This greatly expedited the movement of coca leaves from Peru and other Andean countries by way of "traders" in Bolivia and Ecuador and on to Columbian refineries (40)(41).

The United States has had a continuing problem preventing the transport of drugs from Mexico by autos and trucks. (Should entry be made even easier?) The current status of efforts by Mexican long-distance truckers who want greater access is not clear (42).

HEROIN: Most opium poppies are still grown in Southeast and Southwest Asia. The *Golden Triangle* has traditionally included Burma (Myanmar), Thailand and Laos. The *Crescent Triangle* included Iran as the "key producer", with Pakistan and Afghanistan mostly refining and shipping. The failed Soviet occupation of Afghanistan has been attributed partly to the widespread availability of opium and heroin, with troops raiding their own military supplies to barter them for drugs.

Ironically, the radical Islamic Taliban in Afghanistan were supported largely by opium poppies, despite the conflict with the teaching of the Koran. Considering the tremendous violation of human rights fostered upon a poverty-stricken and starving Islamic populace, this should be no surprise.

Before the U.S. invasion of Afghanistan, the Taliban was able to reduce opium poppy growing by 98%, but in the wake of the war, the poppy harvest is booming again. Farmers still have no alternative means of survival (43).

An enlightening publication from 1972 described an "American heroin empire" built by the worldwide cooperation of corporations, governments, "mafias" and other criminal organizations. Also, it is very worrisome that the CIA was apparently involved in promoting heroin use in southeast Asia during the Vietnam was (1950-1975) (44).

METHAMPHETAMINE:In the U.S., within the past few years, an unexpected, devastating, nationwide epidemic of this very addicting "home-made" drug has been destroying the brains of teenagers and young adults. Oregon in particular was hard-hit. Atlanta, Georgia, has been identified as the distribution hub for Mexican-run "super-labs." A delayed and disjointed federal prevention program was blamed on the pharmaceutical industry's objection to limiting availability of "cold medicines" containing pseudoephedrine, a key element in the manufacture of this deadly addicting drug. A.U.S. House of Representatives Methamphetamine Caucus of 34 states planned to study the problem (45).

The next topic concerns both legal and illegal global "businesses" which produce monumental amounts of money and, therefore, have power and influence over governments. These are not just the *legal* arms dealers, nor the *illegal* drug cartels. They are the various *mafia-like* criminal including the Sicilian-American, the Russian so-called *Red* mafia, the Japanese Yakuza and Chinese Dragon Syndicates.

INTERNATIONAL CRIMINAL ORGANIZATIONS

[REFS 46-59]

Some of these groups trace their roots to age-old secret *mafia-like* organizations, sometime originating of necessity to combat tyrannical governments, but becoming tyrannical themselves. Other are of more recent origin (46)(47)(48)(49).

The Sicilian-American *mafia* would, perhaps, be of the most immediate interest to American readers. Truly a *tangled web of secrecy and deception,* there is reason to believe that they were especially involved with the Vatican Bank and the American CIA during the *cold war* against the communist Soviet Union.

Conspiracies and assassinations have involved highly placed persons in the American and Italian governments as well as powerful business interests. Centuries of links between the Knights Templar, Knights of Malta, and links with some modern Freemasons in government are reviewed in several references (50)(51).

A well researched and very detailed description of the *Red Mafia* in American and worldwide is available. A shorter version appeared earlier in Reader's Digest (52)(53).

A terrifying and detailed account of the Russian *Mafiya* documents their rapid worldwide takeover of organized crime including extortion, gambling and organized prostitution, laundering drug money for the Colombian cocaine cartel (trading heroin from Southeast Asia for cocaine), smuggling arms (including nuclear materials), assassinating rivals (worse than the Sicilians, they kill the whole family), setting up bogus businesses, including *holding companies* in the Channel Islands. They have taken control of Russian banks, clashed with and cooperated with Italy's most powerful mafia (Camorra) as well as the Japanese *Yakuza* (54).

The Japanese *Yakuza* is "today one of the largest and most far reaching crime organizations in the world. With twenty times as many members as the American Mafia, they are changing the face of Japan's 'crime-free' society and reaching beyond. Members (have) elaborate tattoos,(carry) business cards and (wear) gang emblem pins. They cut off tips of their fingers to show devotion to their 'godfather' (55)."

A network of criminal gangs in Japan preceded the end of World War Two by some 50 years. The American occupation forces insistence on action against them met little enthusiasm.

The Prime Minister denied their existence. Some occupation

officials encouraged the gangs and even paid their leaders. Japan was in ruins. Almost every major city was severely damaged by firebombs or nuclear weapons. In Tokyo, one million out of 1.7 million buildings had been destroyed, most by the fire-bombing raids in 1945. Chaos continued for years as Americans tried to rebuild a democratic Japan (55).

Around 1978, public gun battles and repeated assassinations between the large Yamaguchi-gumi gang (12,000 members) and their major rival, The Matsuda-gumi, eventually resulted in an uneasy truce. An "enraged" public had resulted in 1100 police officers rounding up 2000 gangsters, including 518 senior members.

The Yakuza control gambling, prostitution, drugs, loan sharking, gun smuggling, labor rackets and extortion. They have established a kind of truce with the police, often outnumbering them. They have obtained huge non-repayable bank loans and practically control politicians. They have set up drug-smuggling, gun-running and sex slavery operations in Honolulu, Los Angeles, New York and Las Vegas. They have made a "dramatic impact" on Hawaii, investing over $100 million in linking up with the islands most powerful crime figures."

Today they seem to have a place of honor in Japan's "popular culture." There is far more than can even be referred to in this present writing. For anyone interested in more details, see these excellent references (56)(57).

The Chinese Dragon Syndicates or *Triads* probably date back to the Zhou dynasty (1027-221 BC/BCE), when secret groups banded together to depose any "divine" emperor who betrayed his obligation to set a moral and spiritual example. The first secret society to over throw an emperor was probably the Chih Mei ("Red Eyebrows") in 25 AD/CE. The emperor, Wang Mang, was assassinated.

During the next 2000 years, China's history was one of secret societies, revolts and executions of dissenters.

Hundreds of thousands perished. Today, some 60 million Chinese live outside China and comprise some 5 percent of all ethnic Chinese. The criminal Triad enterprises have followed them throughout the world.

The coming of foreign (European) traders and missionaries in the 1800'S brought chaos on a larger scale. The *Opium War* (1839-1842) resulted when European traders introduced a better grade of opium from India or Persia. Opium had actually been introduced by Arabs in the ninth century. This helped open the Chinese market to foreigners.

One writer states that, beginning in 1788, the first Masonic lodges were established in China, under the auspices of the Grand Lodge of England. Also, "the infamous Triad Society began as a Masonic order, with identical rites, jeweled symbols and leather aprons. One was called the 'Order of the Swastika' (58)."

No one religion or many religions seemed to be the cause of repeated violent incidents involving the destroying of temples, killing of priests, and looting. However, there was a period around 1848 when a fanatical believer in Yasoo (Jesus),Liang A-fa, with little understanding of that religion, induced a Triad leader, Hung Hsiuch'uan (1814-1864), to begin a "holy war" against the Q'ing emperor, destroying temples, monasteries, desecrating idols, and massacring monks. Hung had already started the infamous Taiping Rebellion. In 1851 the capital, Nanjing, fell and Hung proclaimed himself the divine emperor. Some 35 million died in this Rebellion and much of China was laid waste. This should all sound familiar, as similar events in the Middle East and Europe have continued since the early Christian millennium.

During the next 200 years the criminal enterprises of the Triad had spread rapidly around the world. The details are appalling, not for the faint-hearted, but available in the 356 pages of the following reference (59).

REFERENCES AND NOTATIONS: CHAPTER EIGHT

THE SHADOW GOVERNMENTS

1) Arendt,Hannah. *The Origins of Totalitarianism.* New York. Harvest Book/Hardcourt, 1948 and 1994.
2) Korten,David. *When Corporations Rule the World.-Second Edition.* San Francisco. Berrett-Koehler Publishers,2001. [See Chapter 8:*Dreaming of Global Empires* and *Localizing the Global System,* pages 279-284]
3) Nace,Ted. *Gangs of America: The Rise of Corporate Power and the Disabling of Democracy.* San Francisco. Berrett- Koehler,2003.
4) Huffington,Arriana. *Fanatics and Fools.* New York. Hyperion, 2004.
5) Glahn,Gerhard von. *Law Among Nations: An Introduction to Public International Law – Second Edition.* New York and
6) Meron, Theodor. *War Crimes Law Comes of Age: Essays.* Oxford. Clarendon Press, 1998.
7) Gutman,Roy;Rieff,David. *Crimes of War: What the Public Should Know.* New York and London. W.W. Norton,1999.[See *Occupation of Territory,* page 263;*Persecution,*page 272; *Protected persons,* pages 295-296; *Refugee rights,* pages 306-309;*Terrorism,*page 351;*Unlawful confinement,* pages 365-368;*Wanton* destruction, pages 372-373; War crimes, categories, pages 374-376]

FEDERAL RESERVE SYSTEM

8) Sutton,Antony. *The Federal Reserve Conspiracy.* Boring, Oregon. CPA Book Publishers, 1995.
9) Marrs,Jim. *Rule of Secrecy.* New York. HarperCollins, 2000.[See pages 64-78, *Federal Reserve System.*]
10) Still,William. *New World Order: The Ancient Plan of Secret Societies. Lafayette, LA.* Hunting House Publishers, 1990.[See pages 49-51]
11) Sutton,Antony. *The Federal Reserve Conspiracy.*Boring, Oregon. CPA Book Publishers,1995.
12) Marrs,Jim. *Rule by Secrecy.* New York. HarperCollins, 2000. [See page 69 re "*the secretive seven.*"]
13) Marrs,Jim. *Rule by Secrecy.* New York. HarperCollins, 2000.[See pages 64-78]

INTERNATIONAL BANKING

14) Carroll,James. *Constantine's Sword: The Church and the Jews.* New York. Houghton Mifflin, 2001.[See pages 432- 433]
15) Ferguson, Niall. *The House of Rothschild: Money's Prophets, 1798-1848.* New York. Penguin Books, 1998.

16) Korten,David. *When Corporations Rule the World.* San Francisco and Bloomfield, CT. Barrett-Kochler and Kumarian Press, 2000.

17) Rockefeller,David. *David Rockefeller: Memoirs.* New York. Random House, 2002.

18) Marrs,Jim. *Rule of Secrecy: The Hidden History That Connects the Trilateral Commission, the Freemasons, and the Great Pyramids.* New York.Harper Collins Publishers, 2000.[See pages 44-54]

MULTINATIONAL CORPORATIONS

19) www.stopcorporateabuse.org

20) Baltimore,David. *Supremacy slipping away.* Los Angeles Times, December 2004.[The author is president of the California Institute of Technology and has won a Nobel Peace Prize]

21) Korten,David *When Corporations Rule on World. Second Edition. Bloomfield, CT 2001.*[See Part III. Corporate Colonialism, pages 121-174]

22) Zinn,Howard. *A People's History of the United States 1492-Present. Revised and updated Edition.* New York. HarperPerennial, 1995. [See page 556]

23) Veon, Joan. *The United Nations' Global Straightjacket.* Oklahoma City. Hearthstone Publishing, 1999,2000.[See Chapter 6]

24) http://drclas.fas.harvard.edu

25) Weiss, Martin. *Safe Money Bulletin, March 2003.* [See www.oprah-goesonline.com]

26) *The Disappearing Dollar.* The Economist. London, UK. December 4-10,2004.

THE BILDERBERG GROUP

27) Marrs,Jim. *Rule by Secrecy.* NewYork.HarperCollins,2000. [Pages 39-44]

28) Ronson,Jon. *Them: Adventures with Extremists.* New York and London. Simon & Schuster, 2002.[Chapter Three: The Secret Rulers of the world]

29) Marrs,Jim. *Rule by Secrecy: The Hidden History That Connects the Trilateral Commission, the Freemasons, and the Great Pyramids.* New York. Harper Collins, 2000.

30) Rockefeller,David. *David Rockefeller: Memoirs,* New York. Random House, 2002.

THE TRILATERAL COMMISSION

31) Marrs,Jim. *Rule by Secrecy,*2000. [See page 22-31]

32) Korten,David. *When Corporations Rule the World* –Second Edition. San Francisco. Berrett-Koehler Publishers, 2001.[Pages 139-142]

33) Sutton,Antony; Wood,Patrick. *Trilaterals Over Washington.* Scottsdale, AZ. The August Corporation, 1979.

COUNCIL ON FOREIGN RELATIONS (CFR)

34) Marrs,Jim. *Rule by Secrecy,*2000.[See pages 32-38]

ARMS DEALERS

35) Ayton-Shenker,Diana;Tessitore,John (Editors). *A Global Agenda: Issues Before the 56th General Assembly of the United Nations.* New York. Rowman and Littlefield Publishers, 2002.

36)Larsen,Susie. *U.S.Arms Sales: Arms Around the World* [Internet site: http://motherjones.com/arms]

37) Pierre,Andrew. *The Global Politics of Arms Sales.* Princeton NJ. Princeton University Press, 1982.

38) Kidron,Michael;Smith,Dan. *The War Atlas: Armed Conflict - Armed Peace.* New York. Simon & Schuster, 1983.

39) Ismael and Ismael. *The Gulf War and the New World Order: International Relations of the Middle East.* Miami, FL. University Press of Florida, 1994.

DRUG CARTELS

40) Inciardi,James. *The War on Drugs: Heroin, Cocaine, Crime and Public Policy:* Mountain View, CA. Mayfield Publishing Company, 1986.

41) Ayton-Shenker,Diana; Tessitore,John (Editors). *A Global Agenda: Issues Before the 56th General Assembly of the United Nations.* New York. Rowman & Littlefield Publishers, 2002.[See pages 38 and 40. Also see *Drug Control and Crime Prevention* on the 57th U.N.Agenda, pages 183-188]

42) Padgett & Grillo.*Cocaine Capital.* TIME. August 25, 2008.

43) Ahmad,Ishtiaq. A Global Agenda: 59th U.N. General Assembly: *Drug Trade Derails Afghan Reconstruction*. New York. UN Association of the USA, 2004.[See page 149]

44) Kunnes,Richard (M.D.) *The American Heroin Empire: Power, Profits and Politics.* New York. Dodd, Mead and Company,1972.

45) Suo,Steve. *East Coast's horror stories reflect new map of meth.* Portland, Oregon. The Oregonian. 12/31/2004

INTERNATIONAL CRIMINAL ORGANIZATIONS

46) Friedman,Robert. *Red Mafiya: How the Russian Mob Has Invaded America.* New York. Little, Brown and Company, 2000.

47) Kaplan,David; Dubre,Alec. *Yakuza: The Explosive ount of Japan's Criminal Underworld.* Reading, Mass. Addition- Wesley Publishing, 1986.

48) Sundara,Miyui. *Yakuza the Japanese Mafia.* See members.tripod.com/~orgcrime

49) Booth, Martin. *The Dragon Syndicates: The Global Phenomenon of the Triads.* New York. Carrol & Graf Publishers, 1999.

50) Marrs,Jim. *Rule by Secrecy: The Hidden History That Connects the Trilateral Commission, the Freemasons, and The Great Pyramids.* New York. HarperCollins Publishers,2000. [See pages 254-259]

51) Brewton,Pete. *The Mafia, CIA & George Bush.* New York.
Spi Books/Shapolsky Publishers,1992.[Chapter 12: *The Mobsters, the Spooks and George Bush*]

52) Friedman,Robert. *Red Mafiya: How the Russian Mob Has Invaded America.* New York. Little, Brown and Company, 2000.

53) Van Atta,Dale. *From Russia With Blood.* Reader's Digest, December 2000.

54) Friedman, Robert. *Red Mafiya: How the Russian Mob Has Invaded America.* New York. Little, Brown and Company, 2000. [See Chapter 11: *Global Conquest*]

55) Kaplan,David; Dubro,Alec. *Yakuza: The Explosive Account Of Japan's Criminal Underworld.* Reading, PA. Addison-Wesley Publishing Company, 1986. [Chapter Two: *Occupied Japan.* ALSO SEE: Talmadge,Eric. *Japan ran brothels for U.S. Gis.* The Associated Press, April 26, 2007]

56) Dower,John. *Embracing Defeat: Japan in the Wake of World War Two.* New York. W.W. Norton & Company, 1999.

57) Nakano, Minoru. *New Japanese Political Economy and Political Reformation. International Bookshop (IBS),* 2005. [NOTATION: This is a recent and scholarly presentation of the remarkable changes in Japan's economy, from a shambles after the WW II U.S. bombings, to becoming the second largest world economy (after the U.S) in the 1990s.During the 1980s the U.S. lost many basic of its basic industries, purchased by Japan. Some observers placed significant blame on U.S. President Reagan (1981-1989). In early 2003,after 10 years of decline, Japanese stocks bottomed out. Banks were considering dumping $500 *billion* worth of U.S. stocks and treasuries.[Source: Dan Ascani's P*rofits*

Without Borders. [ALSO SEE: Choi,Hyconjung. *Japanese Political Economy in the IT revolution Era.* Purdue University, 2005. [ALSO SEE: Crichton,Michael. *Rising Sun.* New York.Alfred Knopf, 1992.(This historical novel provides much detailed information regarding Japan's huge trade surplus and massive investment in U.S. basic industries and land]

58) Marrs,Jim. *Rule by Secrecy,*2000.[See page 243]

59) Booth,Martin. *The Dragon Syndicates: The Global Phenomenon of the Triads,* New York. Carroll & Graf Publishers, 1999. 165

PART FOUR

Self-directed (Cultural) evolution

CHAPTER NINE

CULTURAL EVOLUTION

CULTURAL EVOLUTION OVERVIEW

[REFS 1-5]

The urgent need for the human species to assume responsibility for its own survival and a possible future is discussed here. A most eloquent and scholarly presentation of this concept (Cultural Evolution) may be found in at least one modern textbook on evolution (1).

During some 100,000 years the only dramatic evolutionary changes in our species (*homo sapiens sapiens*) have been *cultural*, not *physical*. In other words, *human nature* has not changed. Our continued *cultural* evolution occurred only through increasing knowledge of the *natural* world and through the methods of science.

The cultural evolution of humans has been discussed in many other scholarly publications (2)(3)(4). There is evidence that much now depends upon the end of all *world-wars*, with the survival of the United Nations – if this is the path to a "New World Order."

The principle of *evolution* seems somehow related to the process of *self-organization* or *self-assembly* that occurs during the evolution of all species of life and, perhaps, of the expanding universe. Evolution has been described as "akin to learning (5)."

CULTURAL EVOLUTION AND NEW WORLD ORDER

[REFS 6-11]

The New World Order theme has been prominent in the history of Freemasonry, the Roman Catholic (Universal) Church and Communism. Pythagoras (c.582-507 BC/BCE), the famous Greek Philosopher, was allegedly the first to predict a "New World Order". (The source of this statement could not be located. Perhaps it was alluded to by Will Durant (6))

A prevailing idea for centuries has been of a utopian and peaceful world, patterned after a mythical kingdom of *Atlantis,* written about by Plato, probably dating to Plato's *Republic* (c.387 BC/CE).

This utopian society has often been referred to as residing on the mystical island of Atlantis, where "all women would belong to all men and vice-versa, with children raised anonymously" and all property would be held in common. Plato's "Atlantis" resembles some theoretical communist precepts (7)(8)(9)(10)(11).

Sir Francis Bacon (1561-1626 AD/CE), lawyer, member of British Parliament, grand chancellor under King James I, knighted by Queen Elizabeth in 1603, has also been considered the true author of "Shakespeare's" plays, the founder of English Freemasonry, the "guiding light of the Rosicrucians" and "much involved in the underground operations of the Knights Templar traditions." In 1627 he wrote *The New Atlantis,* later to be know as America.

In a *New World Order,* candidates for *Rulers of the World,* as rivals or allies, should include multinational corporations, international banks and stock exchanges, global organized criminal enterprises, secret societies (Freemasons and affiliates, e.g., Skull and Bones Order), the rival religions, Roman Catholic (Universal Church), Islam (Khilafah), Communism and, possibly, Zionist Judaism, among others. (See "Rival Rulers of the World," Chapter Four).

AMERICA, THE NEW ATLANTIS

[REFS 12-14]

It seems doubtful that the "Irish hermits" (who left Iceland (Lydveldid island) when the "pagan" Norse people arrived in the late 800s AD/CE), knew of Plato or "Atlantis." Nor is it likely that Leif Erickson knew of the ancient plan for a "New Atlantis" when he established a Norse colony in New Foundland at L'Anse around 1000 AD/CE.

However, some of the Knights Templar who escaped execution by the Roman pope and French king, Philip IV, between 1308-1312, were believed to have sailed from Portugal to "La Merica," possibly to land near what is now Newport, Rhode Island. Some credible archaeological evidence found near the landmark "Newport tower," dating to the fourteenth century, was referred to by the Italian navigator, Giovanni da Verrazano in 1524 as "an existing Norman Villa." It seems almost certain that the Templars would have known of the "ancient secret plan." If they were actually there, they probably perished at the hands of unfriendly natives.

The Knights Templar have long been considered the progenitors of modern Freemasonry and, surely the ideal of a New World Order (12(13(14).

CONQUISTADORS IN AMERICA

[REFS 15-18]

An earlier period in the "conquest" of America (the New Atlantis) is described in considerable detail elsewhere. This includes Columbus' brutal treatment and near extermination of "Indians" in his search for gold for Spain, for Ferdinand II and Isabella I. All he could bring back were native slaves.

Note that 1492 was the year the last Jews were driven from Spain and the Muslims in 1502. The official religion of Spain then became Catholic.

Christopher Columbus or Cristoforo Colombo (1451-1506) from Genoa, Italy, has been widely accepted as the heroic discoverer of America. "What Columbus did to the Arawaks of the Bahamas, Cortes did to the Azetecs of Mexico, Pizarro to the Incas of Peru, the English settlers of Virginia and Massachusetts to the Powhatans and the Pequotes (15)(16)(17)(18)."

PURITANS, PLATO AND THE INDIANS

[REFS 19-21]

There is some further evidence that America was destined to become the "New Atlantis." Consider the Puritans, Plato and the Indians.

In December 1620 AD/CE, when the *Mayflower* arrived at Cape Cod with Puritan-Pilgrims from England, William Bradford soon instituted a program of "communalism." This had been required by the London merchants who financed the trip. Food shortages made him soon realize that this system of owning everything in common was not working. When incentives were provided by assigning land to each family the Colony prospered.

Bradford later wrote: "The vanity and conceit of *Plato* and other ancients... that the taking away of property and bringing (it) in the community...would make them happy and flourishing: as if they were wiser than God. However, (it) was found to breed much confusion and discontent, and retard much employment that would have been to their benefit and comfort (19)."

In the Massachusetts Bay Colony, John Winthrop had used the Bible to justify taking the Indians' land. "Ask of me, and I shall give thee the heathen for thy inheritance, and the uttermost parts of the earth for thy possession." (Psalms 2:8)

In 1636 the Puritans began destroying Pequot villages in reprisal for the murder by Indians of "a white trader, Indian-kidnapper, and troublemaker." Captain John Mason had said that "Massacre can accomplish the same end", inheriting the earth by attacking Pequot warriors. Both William Bradford and Cotton Mather were pleased with "sending Pequot souls to hell (20)."

Similar events had taken place in Virginia's Jamestown, dating from 1607 when Powhatan pled with John Smith regarding the taking of Indian lands by force.

From 1610 through the 1620s, massacres took place on both sides. It was total war. "Not able to enslave the Indians and not able to live with them, the English decided to exterminate them (21)."

AMERICA'S FOUNDING FATHERS

[REFS 22-30]

Much has been written about the influence of Freemasons and the Catholic Church upon the founding of the United States of America and, even, the design of the capitol city, Washington, DC, its monuments and U.S. money (22)(23).

One writer describes at length the problems American Freemasons had deciding between loyalties to the English king George III and the colonial revolutionaries. He describes a number of prominent Freemasons, and attempts to correct some popular misconceptions regarding their roles as "founding fathers (24)."

There is credible evidence that 55 of the 56 signers of the Declaration of Independence in 1776 were Freemasons, but it seems likely that they knew of no "ancient plan" involving America as the "New Atlantis." (Recall that Sir Francis Bacon, a Freemason, did write *The New Atlantis* in 1627 (25)).

The relationships between the Catholic Church and the Freemasons over the centuries have been puzzling. The Knights Templar, the probable forerunners of modern Masonry, were executed as heretics, beginning on Friday, the 13th of April 1307. In April 1738, Pope Clement XII forbade Catholics to become

Freemasons on pain of excommunication.

At the time of the Inquisition's massacre of the heretical Cathars in southern France, beginning in 1209, some Templars were thought to be "Cathars." "Captured documents" had indicated that the Templars had always rejected the "religion of St. Peter (26)."

The Society of Jesus (Jesuits), founded in 1540 by Ignatios Loyola, a soldier-turned-priest, became a militant force against heretics and Protestants. Adam Weishaupt allegedly modeled his *Illuminati* after them. Over time the Jesuits began resisting the authority of the Roman Church and in 1773 the Order was banned, only to be reinstated in 1814 by Pius VII. The role of the Catholic Jesuits in the American Revolution is described in great detail elsewhere (27).

Some authors have said little or nothing about Jesuits in their description of the Freemasons in the American Revolution (28)(29)(30).

NEW WORLD ORDER AND THE UNITED NATIONS

[REFS 31-32]

A number of books have been written and the Internet has many web-sites with lengthy and well-documented concerns regarding what the United Nations may portend for the cherished freedoms of a true democracy (31)(32).

In particular, these include freedom of religion, speech and press, peaceful assembly, the petition of government for redress of grievances (First Amendment to the U.S. Constitution, one of ten in 1791): The right to bear arms for the security of a free state (Second Amendment); freedom from unreasonable searches and seizures (Fourth Amendment); rights in criminal cases and to a fair trial (Fifth and Sixth Amendments).

The most significant Amendments enacted after 1791 include

the abolition of slavery (1865); civil rights (1868); black suffrage (right to vote) (1870); income taxes (1913); women's suffrage (right to vote) 1920). [See Appendices A (U.N. Charter) and B (U.S. Constitution- Bill of Rights).

Of utmost importance are provisions in the Constitution of the United States (1787) which require separate Legislative, Executive and Judicial Branches of the federal government. The concept of "separation of church and state" seems to be included in the First Amendment (freedom of religion, speech, the press and right of petition). There is some reason to believe that some powerful group or groups are attempting to destroy some basic principles of the U.S. Constitution, e.g., "Constitutional conventions".

REFERENCES AND NOTATIONS: CHAPTER NINE

CULTURAL EVOLUTION OVERVIEW

1) Strickberger,Monroe. *Evolution - Third Edition.* Sudbury, Massachusetts. Jones and Bartlett,2000. [See *Culture and the Control of Human Evolution.*]
2) Smith,John Maynard; Szathmary,Eors. *The Origins of Life: From the Birth of Life to the Origin of Language.* New York. Oxford University Press,1999. [CHAPTER 12: *From Animal Societies to Human Societies*]
3) Fox,Ronald. *Energy and the Evolution of Life.* New York. W.F.Freeman, 1988.[See Chapter 3:Self-assembly and Self- Control]
4) Johnson,Steven. *Emergence: The Interconnected Lives of Ants,Brains, Cities and Software* New York.Scribner, 2001. 5)Buckley,Walter (Editor). *Modern Systems Research for the Behavioral Scientist: A Sourcebook.* Chicago. Aldine Publishing, 1968.[See Pringle,J.W.S. *On the Parallel Between Learning and Evolution*]

CULTURAL EVOLUTION AND NEW WORLD ORDER

6) Durant, Will. *The Story of Philosophy.* New York. Pocket Books, 1926.
7) Still, William. *New World Order: The Ancient Plan of Secret Societies.* Lafayette, LA Huntington House Publishers, 1990.[See pages 129 and 152]
8) Marrs,Jim. *Rules of Secrecy: The Hidden History that Connects the Trilateral: The Hidden History That Connects the Trilateral Commission, the Freemasons and the Great Pyramids.* HarperCollins Publishers, 2000. [Pages 228,350]
9) Saussy,F. Tupper. *Rulers of Evil: Useful Knowledge About Governing Bodies.* New York. Harper Collins Publishers, 1999.
10) Marx, Karl and Engels, Friedrich. *Communist Manifesto,* 1848.
11) Veon,Joan. *The United Nations' Global Straightjacket.* Oklahoma City, OK. Hearthstone Publishing, 1999. [Chapter 1: The Hegelian Dialectic ; New World Order]

AMERICA, THE NEW ATLANTIS

12) Still William. *New World Order,*1990. See Chapter Three: *The Great Atlantean Plan.*
13) Marrs,Jim. *Rule by Secrecy.* New York. HarperCollins *Publishers,* 2000. [See *"Sir Francis Bacon and the New Atlantis,"* pages 227-232]
14) Ridley,Jasper. *The Freemasons: A History of the World's Most Powerful Secret Society.* New York. Arcade Publishing, 2001. [See page 288]

CONQUISTADORS IN AMERICA

15) Zinn,Howard. *A People's History of the United States 1492- Present.* New York. HarperPerennial, 1995.[Chapter One: Columbus, the Indians and Human Progress]

16) Zinn,Howard;Arnove,Anthony. *Voices of the People's History of the United States.* New York and London, Seven Stories Press, 2004.[Chapter One for more detailed descriptions from eyewitnesses regarding the genocide of West Indies natives by Spaniards – Columbus especially]

17) Carey,John (Editor). *Eyewitnesses to History.* Cambridge, MA. Harvard University Press, 1987.[*"Spanish Atrocities in the West Indies"* and *"With the Spaniards in Paraguay"*]

18) Paris,Edmond. (Translated from the French). *The Secret History of Jesuits.* Ontario, CA Chick Publications, 1975. [*"The Americas: The Jesuit State of Paraguay."*]

PURITANS, PLATO AND THE INDIANS

19) Still,William. New World Order: The Ancient Plan of Secret Societies. [See pages 57-59; *Plymouth*] [Also see Nathan Philbrick. *Mayflower: A Story of Courage,Community and War.* New York and London. Penguin Group,2006]

20) Carr,Caleb. *The Lessons of Terror: A History of Warfare Against Civilians.* New York. Random House, 2002.

21) Zinn, Howard. *People's History of the United States:1492 to Present.* New York. HarperPerennial, 1980 & 1995.[See Chapter One:*Columbus, the Indians and Human Progress*]

AMERICA'S FOUNDING FATHERS

22) Saussy,F.Tupper. *Rulers of Evil.* Chapter 20: American Graffiti.

23) Still,William. *New World Order* , 2006.[Chapter Four and pages 65-68]

24) Ridley,Jasper. *The Freemasons,* 2001. [Chapter Nine: The American Revolution]

25) Marrs,Jim. *Rule by Secrecy.* New York. HarperCollins Publishers, 2000. See part III: Rebellion and Revolution, pages 232-234, *American Revolution.*

26) Marrs,Jim. *Rule by Secrecy,* 2000. [See Part IV, pages 271-341.]

27) Saussy,F.Tupper. *Rules of Evil.* New York. HarperCollins Publishers, 1999.[See page 38 and Chapters 14 through 18]

28) Ridley,Jasper. *The Freemasons,* 2001.[Chapter Nine: The American Revolution]
29) Still,William. *New World Order.* 1990.[See Chapter Four]
30) Marrs,Jim, 2000. [*The American Revolution,* pages 232-235] New World Order and the United Nations

NEW WORLD ORDER AND THE UNITED NATIONS

31) Veon,Joan. *United Nations Straightjacket.* Oklahoma City. Hearthstone Publishing, 1999, 2000.See page 86: Reinventing the Government in the United States.
32) Still,William. *New World Order: The Ancient Plan of Secret Societies.* Lafayette, Louisiana. Huntington House Publishers, 1990.See pages173-174 on the United Nations and Chapter 15:*The Constitutional Assault.*

THE UNITED NATIONS

PERSONAL PROLOGUE

[NO REFS]

In 1945, when the United Nations Charter was signed in San Francisco, this author was still in the U.S. Army, as a medical student at Stanford University. My firm convictions have continued regarding the absolute necessity for a *United Nations* – if *mankind* is to continue to evolve.

Since then I have learned much about the true nature of humans and their history as well as the nature of life and the universe. Many events have brought about both my present hopes and fears for the future of the United Nations and its importance for the survival of mankind.

Most troublesome is an apparent *tangled web* of deception and contradictions in my own country's political agenda for an allegedly safe and prosperous world for all its inhabitants.

This may be more clear to my readers as they study the previous parts of this book and past and current programs of the United Nations. Major changes in the U.N. may be needed to meet its original humanitarian goals.

The United Nations seems clearly to hold the greatest hope

for human survival in the face of so many nations with weapons of mass destruction and an apparent willingness to use them in global conquest or in resistance to such by others. Therefore, it seems of the utmost importance that "citizens of the world" understand and monitor the United Nations to protect its Charter even as Americans must protect their Constitution.

In late 2004, even the continued existence of the U.N. had seemed endangered. This seemed mostly due to the U.S. need for domination.

The United Nations may hold the most promise for the survival and continued evolution of the human species on a peaceful and livable planet. No other planet will ever be available to our species.

ORIGINS OF THE UNITED NATIONS

[REFS 1-6]

In 1945,the United Nations replaced the ineffective League of Nations, organized in 1918 after WW I. "The United Nations grew out of the fervent desire of the United States and its WW II allies to save succeeding generations from the scourge of war (1)."

During World War I, "influential groups in the United States and Britain promoted the concept of a League of Nations," given top priority at the Paris Peace Conference in January 1919, led by U.S. President Woodrow Wilson. Allegedly, "the French tried to establish a League with its own army, completely led by generals who could move against aggressors without the permission of the member states." The U.S. Congress would not ratify U.S. membership and the League gradually atrophied until it was replaced by the United Nations at the end of World War Two.

In 1912, President Woodrow Wilson's "trusted advisor" know only as Colonel Edward House, had written in his novel "Philip Dru": "America is the most undemocratic of democratic countries

... Our Constitution and our laws served us well for the first hundred years of our existence, but under the conditions of today they are not only obsolete but even grotesque" and "Nowhere in the world is wealth more defiant, and monopoly more insistent than in this mighty republic ... and it is here that the next great battle for human emancipation will be fought and won (2)."

In August 1942, Franklin Roosevelt and Winston Churchill met off the coast of Newfoundland and completed the "Atlantic Charter." This document stressed their post-war goals of "seeking no aggrandizement, territorial or other" and respected "the right of all peoples to choose the form of government under which they will live."

However, in secret, the U.S. government had only recently assured the French they could keep their empire as intact as it was in 1939—including Vietnam (3).

Early in WW II, Secretary of State Cordell Hull is quoted as saying "Leadership toward a new system of international relationships in trade and other economic affairs will devolve very largely upon the United States because of our great economic strength. We should assume this leadership and the responsibility that goes with it, primarily for reasons of pure national self-interest (4)."

This should recall to mind statements by Benjamin Disraeli (c.1884) and Cecil Rhodes (c.1897), quoted earlier under *Multinational Corporations*, as Candidates for Rulers of the World.

In 1941, the Atlantic Charter, as a world peace proposal, was designed by President Franklin Roosevelt and Prime Minister Winston Churchill. In 1942, a "Declaration of the United Nations" was signed by 47 nations in support of the Atlantic Charter.

Detailed plans for the United Nations were made at Dumbarton Oaks (1944) and Yalta (1945) by Stalin, Roosevelt and Churchill.

In San Francisco in April 1945, with 50 nations attending, the

Charter of the United Nations was completed, signed in June and was in force in October (5).

The United Nations Charter can be found in Appendix A. Compare this with the United States Constitution, which appears in Appendix B.

In late 2004, the survival of the United Nations itself was in doubt. The reason for this seemed mostly the newer U.S. policy of *unilateralism,* disregard for treaties and international law (6).

UNITED NATIONS STRUCTURE & AGENDAS

[REFS 7-10]

By 2005,the sheer size (over 5,000 staff) and complexity of the U.N., its affiliated agencies and area treaties was almost too overwhelming to understand and monitor. The General Assembly alone consisted of 189 nations in the year 2002.

Unfortunately, the annual publication by the United Nations Association of the United States of America is no longer published. It was entitled *Global Agenda: Issues Before the General Assembly of the United Nations*. It was used as a reference many times in the 2005 first edition of this book (*One Moment in Eternity – Human Evolution).*

A comprehensive chart of the U.N. structure and Agendas is now available on the is now available on the internet at www.un.org (7).

For those who may be interested in some past and present UN agendas, information is now available on the Internet. They have changed little. Also consider noting the United Nations information in TIME Almanac-2006 (many Index items) and The New York Times *Guide to Essential Knowledge – Second Edition, 2007,* pages 1110-1114 (8).

In the opinion of this present author, it should be of great importance to my readers to know the qualifications of the current Secretary General, Ban KI-moon, Republic of Korea, and the President of the General Assembly, Doctor Srgjan Kerim, of Yugoslavia. Both have long and distinguished diplomatic careers (9).

Current issues in 2007 before the General Assembly (found at www.un.org/issues) included *population, climate change and environment,* terrorism, Palestine and Iraq, *AIDS, women and children, human rights, disarmament, drugs and crime, international law.*

Global Warming has been called a high priority, as "a grave and growing problem", by the new Secretary General Ban Ki Moon [Washington Post, March 1, 2007].The global population "explosion" should be the most urgent underlying problem.

The United States continued as the leading supplier of weapons to the *developing world* in 2006, followed by Russia and Britain. The top buyers were Pakistan, India and Saudi Arabia.[Source: New York Times News Service ,October 1, 2007]

This present author proposes that the greatest of needs could be for all nations to become subject to the same laws which equally protect all world citizens. Again, these are exemplified in the United States Constitution and the United Nations Charter. See APPENDIX A and B.

Therefore, at this point, my readers may have a special interest in reading more about the International Criminal Court (ICC). It could have the same significance to the world as the Supreme Court has to the United States (10}.

INTERNATIONAL CRIMINAL COURT (ICC)

[REFS 11-14]

This organization is now a permanent part of the United Nations main body. On July 17, 1998, at an international conference in Rome, 120 nations adopted a treaty to establish a permanent

International Criminal Court. Seven countries voting *against* the Treaty included the United States, Israel, China and Iraq. There were 21 abstentions (11).

Ratified April 11, 2002, by the required 60 nation vote, this new treaty was quickly denounced by President George W. Bush, U.S. State Department staff and Defense Secretary Rumsfeld. Although it had been signed on December 31, 2000, by former President Clinton, it was no longer considered to be legally binding. It was also said to relieve the U.S. of obligations under the Vienna Convention on the Law of Treaties (1969). The new Court supposedly would cede U.S. sovereignty to an international prosecutor who could initiate "capricious and politicized" charges against U.S. officials and military officers. Widespread criticism of the Administrations' position came from a number of human rights organizations, and was expected to further harm American credibility and aggravate some closet allies (12)(13).

The ICC had been ratified by 77 countries by August 2002. It is charged with investigating and prosecuting individuals accused of crimes against humanity, genocide and war crimes. It will become involved where the national courts are unwilling or unable to do this.

Prior to this new permanent Criminal Court, ad hoc criminal Tribunals served this function. These included the International Criminal Tribunal for the Former Yugoslavia (ITY)and the International Criminal Tribunal for Rwanda (ICTR). The history of war crimes law is presented at length elsewhere (14).

REFERENCES AND NOTES: CHAPTER TEN

PERSONAL PROLOGUE

ORIGINS OF THE UNITED NATIONS

1) *What Every American Should Know About the United Nations.* New York. United Nations Association of the United States of America.

2) Still,William. *New World Order,*1990. [See page 156 and a reference to Carroll Quigley, *Tragedy and Hope,*1966]

3) The *Top Secret Pentagon Papers* (Some 7,000 pages).The New York Times, 1971.

4) Zinn,Howard. *A People's History of the United States.* New York. HarperCollins, 1980 and 1995. [Page 403]

5) www.un.org/aboutun/charter

6) *The United Nations Fighting for Survival.* The Economist, November 20, 2004.

UN STRUCTURE AND AGENDAS

7) Drakulich,Angela (Editor).*A Global Agenda: Issues before the 59th General Assembly of the United Nations.* New York. United Nations Association of the USA, 2004.

8) www.un.org [ALSO SEE: *UNITED NATIONS* Handbook, 2007/2008 – 55th Edition. Available through www.ipd@gov.nz]

9) TIME Almanac–2006. Boston,MA .Time,Inc. [See *United Nations* Hist ory,*Structure,Missions and Members,* pages 682,907-909]

10) www.un.org/sg/biography.shtml(Ban Ki-Moon, Secretary General. ALSO: H.E Dr.Srgjan,President,UN General Assembly. www.un.org/ga/president/62/presskit/president.shtml

INTERNATIONAL CRIMINAL COURT (ICC)

11) Aton-Shenker,Diana; Tessitore,John (editors). *Global Agenda: Issues Before the 56th General Assembly of the United Nations, 2001-2002 Edition.* New York. Rowman & Littlefield Publishers. [See Part IV: Legal Issues, pages 239-294]

12)Nye,J. *The new Rome meets the new barbarians.* London. The Economist, March 23,2002.

13) Lewis,Neil. *U.S.refuses help to international court.* New York Times News Service, May 5, 2002.

14) Meron,Theodor. *War Crimes Law Come of Age.* Oxford. Clarendon Press, 1998.

PART FIVE

Summary and conclusions

CHAPTER ELEVEN

AUTHOR'S SUMMARY

Evolution is a Great Moving Force (GMF), operating at all levels from supra-galactic to sub-atomic. The evolution of the Universe, of our own galaxy and star, our planet Earth and all life, including the human species, was described in some detail in PART ONE: EVOLUTION.

Modern humans first appeared only about 100,000 years ago and have not changed in physical appearance, nor in their instinctive behaviors. Such *physical* changes are expected to occur only over hundreds of thousand of years (DNA mutations).

Continued human evolution must now be self-directed or *cultural,* through a learning process. Human *instincts* drive behaviors and have not changed, and cannot, within any foreseeable future. The most basic instinct for all living things is *survival,* individually *and* as a species.

Written history of the human species (dating from around 4000-3000 BC/BCE) indicates almost continuous wars of conquest accompanied by the slaughter of innocents (on both sides), looting, burning, torture, rape and the taking of slaves.

It is truly amazing - the amount of *Cultural* evolution that has actually occurred in these 100,000 years, and recently at an *accelerating* rate. This has occurred within the same human *physical* makeup, including the same brain, and has been motivated by the same primitive survival instincts. This *cultural* evolution has been a *learning* process. It has a long way yet to go.

The process of cultural evolution is by the accumulation and application of new knowledge of the *natural* world. This is done through the application of the scientific method, by discovering and understanding the *real* world, including ourselves. Human conquest of diseases and control over hazardous environs has *not* been through self-proclaimed prophets who convinced followers that they spoke for one or more imaginary *gods.* Many of the most savage and brutal crimes against humanity have been committed "in the name of God or Allah," whereas the true motives were greed and power.

Therefore, it must not be surprising, that this primitive behavior continues, despite the monumental advances in science and technology (e.g., control of pandemic diseases, and space travel, nuclear power, computers, and the world-wide Internet).

An *ancient Plan* (after Plato), for a *new world order,* to be established on a peaceful and healthy planet, might be possible. It could come about partly through the power of American military, money and media. When thought of as a PAX AMERICANA, it calls to mind the PAX ROMANA, beginning with Octavian (a.k.a. Augustus) from around 30 BC/BCE and continuing to about 200 AD/CE (Dates vary somewhat on precisely when the Roman Republic was replaced by an Empire). In our own near future, a new *global* empire (or Order) might be safely accomplished only through the United Nations, but without perverting the democratic principles of the United Nations Charter. The greatest concern for some *world citizens* is that this *New World Order* could become a global dictatorship by a relatively few secretive multibillionaire "elite" individuals, where the freedoms guaranteed by the U.S. Constitution and the U.N. Charter would be rescinded.

Also frightening is the prospect of a *communistic* society, as proposed by Plato and Marx. This was historically shown to be unworkable in Puritan New England and in Stalin's Soviet Union and, in a way in Hitler's Third Reich. A *utopia* cannot be created upon the dead bodies of thousands or millions of innocents. History shows that hatreds, thirsting for revenge, can persist for hundreds of years.

Those historic facts have been well summarized as follows: "Empires have risen and declined. Wars have been fought for the sake of power, resources or vengeance. Religion has brought out the best in mankind, and the worst (1)."

There is considerable evidence that very powerful entities continue to either compete or secretly cooperate to try and become the "Rulers of the World." Some hide behind religion as their motivation, when it is actually *lust for power.* Others act as if they really have no "religious" (perhaps meaning humanitarian) beliefs at all. They could more be trusted if they could be more easily identified. Recall how Cicero (106-43 BC/BCE) preferred the obvious enemy, openly displaying their banners *outside* the city gates.

Those who control the media, the military and the world's banks could actually "control the world," were it not for the freedom of speech and access to information now provided by a mostly uncensored electronic Internet. This communication medium (the Internet) has created a major turning point in human history, far beyond the printing press. It provides an *information explosion,* at least for those who have access to it.

Shocking as it could sound, a legally elected U.S. Congress, if not too ignorant and/or corrupt, may be the only hope for avoiding a world dominated by an *empire* of tyrannical billionaires with a *mafia-mentality.* In this book, they have been referred to as *shadow governments.* There are also many very powerful *openly* criminal organizations operating on a global scale. The promise of democratic, peaceful and prosperous world through the United Nations is in grave danger. The future for humanity and natural resources seem to mean nothing to those who are in the process of looting and destroying the environment.

This prospect of a new Anglo-American global empire seems to call for a review of some of the *historic empires*, their creation, duration and fate. All were created by force of arms and at great cost. Hopefully, many of my readers have already studied them.

It would be most appropriate to begin with the Empires of *Mesopotamia*:

- Assyrian Empire: Created 2500 BC/BCE and conquered in 612 BC/BCE by Persians. This long period was characterized by "brutal rule."
- Persian Empire: From 612 BC/BCE until Alexander the Great vanquished Darius III in 331 BC/BCE.
- Alexander the Great (356-323 BC/BCE: Ruled an empire from 331 BC/BCE until his death in 323 BC/BCE.
- Seleucius I (358-280 BC/BCE: Alexander's general who then ruled the Macedonian Empire until about 140 BC/BCE with the Parthian conquest.
- The Roman Empire was *established* under the emperor Octavian/Augustus (ruling 27 BC/BCE – 14 AD/CE), and "ended" — or "evaporated" — by 410 AD/CE (2).

Mesopotamia was ruled from 165 AD/CE by the Romans, followed by Persians, Arabs, Mongols, Persians (again), then Ottomans until 1918 AD/CE, with the end of World War Two. *Mesopotamia* then became Syria (under France); Palestine (under the British); Lebanon (under France); Iraq (as a British mandate from 1920, then declared a kingdom in 1922 and a Republic in 1958. Iran, and Jordan (formerly Trans-Jordan) became a British mandate, then independent in 1923.

The *Holy Roman Empire* had replaced the original Roman Empire when the Catholic popes, took the place of Roman emperors by appointing them. This began in 800 AD/CE, when Charlemagne was crowned emperor by the Pope in Rome. The last Holy Roman Emperor was Franz Joseph I, who died in 1916.

In 1804, France was proclaimed an empire by Napoleon I (Bonaparte), who declared himself emperor. This empire ended in 1815 with the restoration of a monarchy. Then it was again declared an empire, for the second time, by Napoleon III (Louis Napoleon), lasting as an empire until 1870.

A Russian *empire* actually began with the first czar, Ivan IV, the "terrible" (1533-1584) lasting until 1918, when the last czar, Nicholas II, and his family were murdered by "revolutionists".

The British Empire, "the greatest empire the world had ever known" and "the Empire on which the sun never set," included India by 1871,known as "the jewel in the crown." The beginnings of this empire are somewhat unclear, but the British had established the East India Trading company in 1600, followed by the first colony in America in Virginia in 1607. They then proceeded to establish their rule in the Caribbean (1625-1655); claimed the continent of Australia after the arrival of Captain James Cook in 1770, followed by the settling of six colonies (1786-1859) and, finally, by a federation of the Commonwealth of Australia (1901). New Zealand was "annexed" in 1840. The Union of South Africa became part of the Commonwealth of Nations in 1910. In 1882, the British occupied Egypt, made it a "protectorate" in 1914 and a sovereign state (Arab Republic of Egypt) in 1922.

In 1776, Britain had lost the "American" colonies, but then established holdings throughout Asia, Africa, Central and South America, and in the Indian, Pacific and Atlantic Oceans.

So much space has been given to this "most global empire" because it did not end from any inner collapse or by falling to a foreign invader. A critical difference seems to be that British conquest of others was followed by establishment of freely elected democratic forms of government, with basic protections similar to those that were provided by the U.S. Constitution and even the English Magna Carta of 1215.

Some vital lessons from history may help prevent another "world empire," accomplished only by force of arms, whether Anglo-American or with assistance from many nations. History could help prevent the possibility of self-destruction, as with most historic empires (3)(4)(5)(6).

REFERENCES AND NOTATIONS: CHAPTER ELEVEN

1) Lewis,Brenda (General Editor). *Great Civilizations.* Bath, U.K. Parragon Publishing, 1999.[*See* page 6 and *Mesopotamia,* pages 196-211]

2) Stearns,Peter (General Editor). *The Encyclopedia of World History – Sixth Edition.* New York. Houghton Mifflin, 2001. [*See The Roman Empire,* pages 75-99]

3) Gibbon,Edward. *The History of the Decline and Fall of the Roman Empire.* [First published 1776-1788. Edited by David Womersley. New York and London. The Penguin Press, 1944]

4) Garraty,John;Gay,Peter (Editors). *The Columbia History of the World.* New York. Harper & Row, Publishers, 1972. [Chapter 19: *The Augustine Empire*]

5) Stearns,Peter;Langer,William. *The Encyclopedia of World History – Sixth Edition.* Boston and New York. Houghton Mifflin, 2001.[*See* Rome pages 75-99 with the collapse of the Roman Empire by 461 AD/CE]

6) Soros,George. *American Supremacy.* New York. Public Affairs-Perseus Books, 2004.

CHAPTER TWELVE

AUTHORS CONCLUSIONS WITH DEGREES OF CERTAINTY

Based on the evidence reviewed here, all facts may be considered in terms of probabilities. This author's current conclusions are listed here as in the legal terminology of judges and juries, referred to as the degree of certainty or proof: Beyond a Reasonable Doubt, or 90 percent certainty; Clear and Convincing, 70 percent or more; and Preponderance of evidence (More-likely-than-not) which requires more than a 50 percent probability.

In a sense, this approach is to determine the extent to which *logic* or *emotion* are most involved – with *intuition* somewhere between.

To seek the truth about the possible future of their species should be the right and duty of every world citizen.

EVOLUTION

[AUTHOR'S CONCLUSIONS]

BEYOND A REASONABLE DOUBT (90 PERCENT CERTAIN)

1) The planet Earth is round (approximately) and is not the center of the Universe.(Remember Giordano Bruno, burned alive in 1600 AD for insisting otherwise, defying the Pope.)

2) Evolution is as certain as "the world is round." (The overwhelming scientific evidence is, of course, not to be found in any of the *Holy* books).
3) What is now known regarding the nature of humankind and the universe is only the *tip-of-the-cosmic-iceberg.*

HUMAN NATURE: BASIC INSTINCTS

[AUTHOR'S CONCLUSIONS]

BEYOND A REASONABLE DOUBT (90 PERCENT CERTAIN)

1) Humans are undoubtedly the most cruel and destructive, but also the most compassionate and creative of all animals on this planet.
2) The two *Basic Human Instincts* are for Self-survival and for Species-survival. The *S*elf-survival instinct accounts for the historical human denial of the permanency of death. The Species-survival instinct is manifested by the powerful need for sexual relations between males and females – as with all animal species.
3) The human belief in a personal *life-after-death* is a product of the *personal survival* instinct. All animals have a *personal survival instinct.* Only humans differ by having tremendous powers of imagination. In the absence of enough solid scientific evidence regarding their own *origins and fates*"(as well as those of everything else in the universe), explanations have historically been *revealed,* usually through angels to self-proclaimed prophets. These contrived theories are often held as unquestionable and unchangeable facts. These views differ completely from the never-ending scientific search for understanding.
4) A number of very destructive human behaviors do not seem explained by either of these two basic instincts alone. They

were discussed earlier under *Self-extinction Behaviors.* They include the historical status of women as virtual slaves to men, and the many destructive behaviors now classified as *crimes-against-humanity.* These continue to include taking of slaves, rape and torture, human *trafficking,* child prostitution, female genital mutilation (FGM).

5) What is now known regarding the nature of mankind and the universe is only the "tip" of the (cosmic) iceberg .

SELF-EXTINCTION BEHAVIORS

[AUTHOR'S CONCLUSIONS – CONTINUED]

BEYOND A REASONABLE DOUBT (90 PERCENT PROBABILITY)

Survival of the human species (or of civilization-as-we-know-it) is in grave doubt due to several basic problems and events, historic and current, as follows:

1) The profound ignorance of most humans regarding the scientific evidence for the three *Conclusions* listed above under *Evolution* , as *Beyond a Reasonable Doubt.*
2) The global *population explosion,* resulting in the destruction of the planet's resources, is comparable to a cancer destroying its host and, thereby, itself. This continuing trend is more dangerous than the continuing prospect of a nuclear holocaust or a catastrophic, incurable, global disease pandemic.
3) The increasing competition for the planet's natural resources by multinational corporations – is often under the guise of free trade and promoting *democracy.* Historically, this has been by colonization, but since WW Two, often by the physical invasion of other countries – based upon deception.
4)The increasing global presence of militant religious fanatics.

They require "converts," needing them to sustain their own beliefs, unfounded in solid physical evidence. They include those who still believe in "death-to-infidels", as in the historic *holy wars*, crusades, inquisitions and witch-hunts.

5)The ongoing Middle-East *time-bomb*, the Judeo-Islamic Civil War between Abraham's children, has been prolonged by the unprovable belief that God (Yahweh) promised Palestine to the Jews and another God (Allah) promised it to the Muhammadans.

BY CLEAR AND CONVINCING EVIDENCE

(70 PERCENT OR BETTER PROBABILITY)

1) There is a "tangled web of secrecy and deception" involving a number of *Candidates for Rulers of the World*. Their relationships as allies or rivals is still far from clear. They often seem to change – as needed.
2) Priests and politicians, motivated by "prophets" or "profits," continue to prey on people's ignorance and fears, Some would-be "rulers of the world" believe that most of the "peasants" (hourly wage-slave tax-payers) are intellectually incapable of participating in planning the world's future. These would-be *rulers* act as if a "glorious end" justifies any means. Under a centralized elite group, given enough money, the media and the military could achieve their goals.
3) Freedom of religion must mean "separation of church and state." Otherwise, there can be no true democracy. Any nation, if not the entire world, could become a "theocracy," ruled by a single religion. It would mean the demise of science and the end of reason. [See Americans United for Separation of Church and State: www.au.org]

BY A PREPONDERANCE OF EVIDENCE

(MORE-LIKELY-THAN-NOT OR 51 % CERTAIN)

1) More-likely-than-not, the continuing global conflict is really to capture the "minds" of the world's citizens, meaning their understanding of their place in the world and in the universe — a universe so vast no one can grasp its significance, not yet the scientists and, certainly not those who look to ancient imaginary gods to reveal it.
2) The long continued threat of an overthrow of the government of Saudi Arabia by militant Muslims (who oppose the rulers' close ties to U.S.) makes the availability of Iraqi oil an extremely high priority. The Saudi rulers, long indebted to the fanatical Wahhabis, must continue to finance the madrassas (religious schools) where children are taught to hate Jews, Westerners, and Americans - as infidels deserving of death. The extreme wealth of the Suadi princes is a *twoedged sword. T*he whole Middle East could explode if a major uprising against them took place. The use of nuclear weapons in these *holy wars* continues to be a possibility. (Keep in mind the probability that "Western civilization-as-we-know-it" absolutely requires unlimited amounts of oil for the fore-seeable future.

SEVEN MAJOR TIME-BOMBS

[AUTHOR'S CONCLUSIONS – CONTINUED]

BEYOND A REASONABLE DOUBT (90 % CERTAIN)

1) On Earth, no living species has had such potential for self-extinction as homo sapiens sapiens.
2) Of some seven extinction *time-bombs,* the most critical is the global *population explosion*. Ironically, most of the current threats to human survival are secondary to this – and

this alone can probably only be controlled by humans.

3) Devastating *natural disasters* have occurred repeatedly and long before the appearance of the human species or their closest ancestors. Massive extinctions of many life forms have included repeated reversals of the planet's magnetic poles during the past 500 *million years and* major climate changes possibly related to massive meteor strikes some 250 *million years* ago and again between 135 and 70 *million years* ago. Ice Ages have occurred 700 *million years* ago, 500 *million* years ago, 280 *million years* ago and most recently, beginning two *million years* ago, then receding around 10,000 years ago. We are still between Ice Ages. These glaciations have appeared rather suddenly and were correlated with magnetic pole reversals. There is an extremely urgent need to foresee these historic disasters and have extensive plans for coping. [See www.iceagenow.com/Magnetic_Reversal_Chart.htm]

4) Global disease pandemics have killed large percentages of the human population during the *brief* 100,000 years of the (modern)species existence. Perhaps the most recent occurrences include the rapid global spread of *Spa*nish influenza in 1918 and the current mounting deaths from AIDS through 2003, estimated at 20 million.

CHAPTER THIRTEEN

READER'S CONCLUSIONS – WITH DEGREES OF CERTAINTY

My readers are now invited to consider their own conclusions regarding the probability or certainty of fact in regard to the materials presented in the previous pages. This material can be regarded as evidence to be evaluated in the form used by courts of law in the United States.

The most likely to be true is called "beyond a reasonable doubt" (90% certain or better). Next is "clear and convincing" (70% or better). The least certain is called "more-likely-than-not" or by the "preponderance of evidence" (51% or better). The evaluator may also decide that a certain material presented as *evidence* does not qualify as *fact* at even the lowest level of legal certainty and is to be "rejected at this time.

The evaluator may use the symbols BRD (Beyond a Reasonable Doubt), or CC (Clear and Convincing, PoE (Preponderance of Evidence) or RJ (rejected).

EVOLUTION

1) Evolution is a basic Natural Law evident at all levels from supra-galactic to subatomic.

Probability as fact: _____

2) Astronomers and astrophysicists have conclusive evidence that stars, like our own sun, are born, evolve and die.

Probability as fact:____

3) All life on Earth has evolved over some 4 billion years from the earliest primitive molecules, then to cells and then multi-cellular plants and animals, only very recently including mankind.

Probability as fact:____

4) Modern humans are an animal species, Homo sapiens sapiens, with reliable scientific evidence of their first appearance around 100,000 years ago.

Probability as fact:____

5) The dating of the death of animal and plant fossils by measuring the remaining amounts of radioactive carbon or potassium-argon is acceptably accurate.

Probability as fact: _______

Key references for reader's possible review:

- Strickberger, Monroe. *Evolution – 3rd Edition,*2001.
- Shklovskii I.S. and Sagan, Carl. *Intelligent Life in the Universe.* Dell, 1982.
- Chaisson, Eric. *Cosmic Evolution: The Rise of Complexity.* HarperCollins.

HUMAN NATURE

[READER'S CONCLUSIONS]

1) There has been no significant physical change in the brain of modern homo sapiens sapiens, nor in instinctive behaviors. This will probably continue for many more thousands of years of their physical evolution.

Probability as fact: ________

2) The only profound changes in the (modern) human species have been through *Cultural Evolution,* which is a learning process.

Probability as fact: _______

3) The Two most basic human instincts are: Self-survival and Species survival.

Probability as fact: _______

4)The Self-survival instinct – present in all animals – is responsible in the human species for the oldest, most persistent and passionate belief in some kind of l*ife-after-death.* In humans this is possible due to the unique ability to imagine possibilities before they are proven. Consider that men have created *all* the gods.

Probability as fact: _______

Key references for reader's possible review:

- Strickberger,Monroe. *Evolution – 3rd Edition.* 2000.
- Armstrong,Karen. *A History of God: The 4,000-year Quest of Judaism, Christianity and Islam.* Ballantine Books, 1993. *See* page 233 re human imagination and gods.
- Berra,Tim. *Evolution and the Myth of Creationism: A Basic*
- *Guide to the Facts in the Evolution Debate.* Stanford University Press, 1990.
- Garraty,John;Gay,Peter (Editors). *The Columbia History of the World.* New York. Harper & Row, 1992. [See Chapter 8: *Gods and Men*]

HUMAN SELF-EXTINCTION: SEVEN TIME-BOMBS

[READER'S CONCLUSIONS – CONTINUED]

1) There is a strong possibility that humans now have the power to bring about their own extinction as a species or, at least, to end civilization as we now know it.

Probability as fact:_____

2) Ironically, the human population "explosion," by itself, may be the greatest danger to human survival.

Probability as fact:_____

3) There is credible evidence that several powerful groups, as rivals or as allies, have historically sought world domination, including the Catholic (Universal) Church, the Islamic Mujahiddin, the international corporations and bankers and, possible, secret organizations such as the Freemasons. Some march under the banner of "religion," others under that of a "New World Order."

Probability as fact: ____

HUMAN SELF-DIRECTED (CULTURAL) EVOLUTION

[Reader's Conclusions – Continued]

1) Humans may direct their own future in a peaceful and prosperous world - for all its citizens - by making available the "explosion" of scientific knowledge and by making enforceable international laws to prevent "crimes against humanity."

Probability as fact:_____

Key references for reader's possible review:

- Maddox,John. *What Remains to be Discovered: Secrets of the Universe, the Origins of Life, and the Future of the Human Race.* New York. Touchstone-Simon & Schuster, 1998.
- TIME Almanac-2006. See *Ancient Civilizations,* page 664.
- Lewis,Brenda (General Editor). *Great Civilizations.* Bath, U.K. Parragon Publishing, 1999.[Especially *see* page 6]
- Gibbin,Edward. *The History of the Decline and Fall of the Roman Empire.* First published 1776-1788. Edited by David Womersley. New York and London. The Penguin Press, 1944.
- TIME Almanac-2006.See *Great Disasters,* pages 210-216, and AIDS, pages 555-557.

PART SIX

Epilogue

CHAPTER FOURTEEN

EPILOGUE

HUMAN SPECIES – PRESENT AND FUTURE

[REFS 1-9]

An Epilogue is intended to provide an update on significant events as near as possible to the date of this publication — in this case, late 2009. A chronological approach will continue where possible.

Again, this essay was intended to provide some of the most basic and vital information about the human species as well as the author's personal opinions regarding what is *delusion* and what is *reality.* Hopefully, it will inspire some readers to pursue their own further research, valuing t*ruth* above *comfort* — the method of scientists.

Following the re-election of George W. Bush as U.S. President in November 2004, and at the time of some of this writing, citizens of the United States continue to be divided as never before since their Civil War - some 150 years ago. At that time the issue was about *greed* (slavery). Today the issue is, again, *greed* — and a possible plan of American corporations to *rule the world* – with or without help from the United Nations or the *shadow governments* described earlier.

The March,2003, U.S. invasion of Iraq for oil precipitated the current great division. It was justified on the pretext that Saddam Hussein was a threat to the U.S. due to his development of weapons of mass destruction — an allegation later proven untrue. Iraqi oil reserves were second only to those of Saudi Arabia, where Saudi princes have been threatened with overthrow by militant Muslims. True, that an unlimited amount of oil is vital to *civilization-as-we-know-it* in the Western world.

No one in the duly elected U.S. Government has yet adequately explained the source of the trillions of dollars to pay for the Iraqi *occupation.* Is this some kind of classified information? (All modern wars were fought on borrowed money.)

During 2008, the *global war on terrorism c*ontinues. Despite the effort to establish a democratic form of government in Iraq, the American body-bags keep coming home. The 4,000 American deaths are small in comparison with Iraqi deaths – due mostly to the continuing Sunni-Shia "civil war" (since Muhammed's death in June 632 CE/AD). The militant Muslims' atrocities, especially against innocent civilians, continue despite some leading Islamic clerics denouncing them as contrary to the Koran.

The apparently illegal *occupation* of Iraq by the United States and a few allies continues. The recently resigned Bush- appointed U.S. Attorney General, Alberto Gonzales, had been quoted as calling the Geneva Conventions *quaint and obsolete.*

Is the continued occupation by Israel of captured Palestinian territories quaint and obsolete (1)(2)(3) ?

Three very important publications have recently shed additional light on the historic causes of *terrorism* and detailed analyses of a hypothetical "better U.S. response" to the massive hijacked-airliner attacks of September 11,2001 (4)(5)(6).

On November 2, 2004, when George W. Bush was re-elected president of the United States, he promptly proclaimed his "clear mandate" for an "aggressive foreign policy." At least half of American voters and most of the rest of the world (especially the Islamic world) could "tremble" as did Abraham and "the Republic (could) be destroyed."

Any Iraqi attempting to vote on January 30, 2005, for a transitional Parliament, risked being killed by the Sunni "insurgents." Some were. The majority Shia and the never-before represented Kurds "won" the election, but Saddam's minority Sunnis continued to protest. Meanwhile, the U.S-led occupying forces seemed unable to protect their own troops. A great many more Iraqi citizens have been killed by *insurgents.*

Although official U.S. statements emphasize that this is not a war against Islam, there seems no other way to view it. "Death-to-infidels" has been a war-cry of Islam for many centuries. As long as this continues — and it will for any foreseeable future — there is no way "infidels" can occupy a Muslim country without the cost of many lives. Most will be Muslim but many will be American.

Elsewhere in the world, the United Nations, now led by Secretary General Ban Ki-Moon, continue to struggle with the chaos of civil wars, crimes against humanity, genocide, and the ever-widening gap between the status of the rich and poor nations - among a multitude of truly global problems.

Early in 2008, the rampant pillaging, rape and genocidal murder of "rebel" villagers in Darfur by Militant Muslim Janjaweed continues. This has been with the assistance of the Sudanese military. Sudan is 70% Sunni Muslim. The "western" world has largely "looked the other way."

. The seventh secretary General of the United Nations, Kofi Annan, in office since 1997, winner of the Nobel Prize for Peace in 2001,had been under assault by U.S. "neo-conservatives" for alleged *corruption* involving Saddam Hussein's personal profit from oil sales not authorized by the Oil-for- Food program of the

United Nations. Any well informed person could strongly suspect that this was *pay-back* time for his opposition to the 2003 U.S. invasion of Iraq by the presidential administration of George W. Bush. Several of the many informative sources on this subject follow here as references (7)(8)(9) .

On December 12, 2004, a tsunami flooded beaches and islands over some 3,000 miles of eleven countries bordering the Indian Ocean following an extremely large underseas earthquake - Richter scale of 9 – with the maximum possible being 10. (Our planet is very fragile !) Indonesia, with a population of over 228 million, is 87 percent Islamic. More than 150,000 deaths were estimated. The global response in terms of aid was overwhelming. After some 3 days, President Bush promised $35 million. Other countries had quickly pledged large sums: the UK, $95 million; France $57 million; Sweden $76 million and Spain $65 million. India, Japan and Australia were forming an aid consortium with the United States.

There had been speculation that the U.S. may have missed an opportunity to gain friends in the Muslim world by their delay in responding. However, more long-term aid in rebuilding towns and villages was promised by the U.S. government (as with the 1847-1951 U.S. Marshall Plan to assist the rebuilding of Europe after World War Two). Massive donations almost overwhelmed aid workers' ability to reach those in the greatest need. Meanwhile, an American aircraft carrier provided helicopters to carry supplies to remote islands.

The Indonesian government set a strict time-line for this to cease. Remember the big picture: Planet Earth is our only home. It is very fragile. No similar planet will ever be within our reach. We are between Ice ages. Historically, there have been *rapid* changes from a warm planet to *full-blown* glaciations in less than 20 years, corresponding to magnetic pole reversals. [See htpp//www.iceage-now.com/Magnetic_Reversal_Chart.htm]

The ultimate fate of our sun's planets is *vaporization* – when our sun becomes a Red Giant.

Key reference for reader's possible review:

- Meron, Theodor. *War Crimes Law Comes of Age.* Oxford. Clarendon Press, 1998.[See Geneva Conventions Chapters VII and VIII][Israeli "occupied territories"]
- Carter,Jimmy. *Palestine:Peace,Not Apartheid.*New York and London. Simon & Schuster,2006. [Also see: *Issues before the 59th General Assembly,*2004-2005. Pages 72-73, *Israeli "occupied territories.*]"
- *The 9/11 Commission Report: Final Report of the National Commission on Terrorist Attacks on the United States.* New York and London. W.W. Norton & Company, 2004.[*See* Chapter Two: Foundation of the New Terrorism. Perhaps the most detailed and concise historical background of "terrorism" available to date].
- Anonymous. *Imperial Hubris: Why the West is loosing the War on Terror. Washington, DC. Brassey's, Inc., 2004]*
- Soros George. *The Bubble of American Supremacy: The Costs of Bush's War in Iraq.* Cambridge, MA. Perseus Books Group. New York. Public Affairs, 2004.
- McGeary,Johanna. *The Fight of his Life.* TIME magazine, December 13, 2004.
- Williams, Ian. *The Right's Assuault on Kofi Annan.* New York, NY. The Nation, January 10-17, 2005.
- Lederer, Edith. *Annan refuses calls to step down.* The Associated Press, 12/08/04. [Many states, as members of the U.N., including Tony Blair, had strongly supported Annan despite George W. Bush's denouncement] 196

APPENDIX A

UNITED NATIONS CHARTER

The Charter was signed in San Francisco on June 26,1945, by 50 nations. By September 2000, membership had increased to 189 nations (1).

"We the peoples of the United Nations, determined to save succeeding generations from the scourge of war, which twice in our lifetime has brought untold sorrow to mankind, and to reaffirm faith in fundamental human rights, in the dignity and worth of the human person, in the equal rights of men and women and of nations large and small, and to establish conditions under which justice and respect for the obligations arising from treaties and other sources of international law can be maintained, and to promote social progress and better standards of life in larger freedom, and for these ends to practice tolerance and live together in peace with one another as good neighbors, and to unite our strength to maintain international peace and security, and to ensure, by the acceptance of principles and the institution of methods, that armed forces shall not be used, save in the common interest, and to employ international machinery for the promotion of the economic and social advancement of all peoples, have resolved to combine our efforts to accomplish these aims.

Accordingly, our respective Governments, through representatives assembled in the city of San Francisco, who have exhibited their full powers found in good and due form, have agreed to the present Charter of the United Nations and do hereby establish an international organization to be known as the United Nations."

Four Amendments to the Charter have been made: Articles 23,

27 and 61 were amended in 1963 and came into force in 1965. An amendment to Article 23 enlarged the Security Council from 5 to 15. Article 27 clarified how decisions of the Security Council were to be affirmed. In 1973, amended Article 61 enlarged membership in the Economic and Social Council from 27 to 54. An amended Article 109 in 1968 described how dates were to be set for reviewing the Charter (2).

References and Notations: United Nations Charter

- World Almanac and Book of Facts – 2001. See page 865.
- Charter of the United Nations and Statute of the International Court of Justice. Published by the United Nations Department of Public Information, October 2000.

APPENDIX B

UNITED STATES CONSTITUTION BILL OF RIGHTS

(First Ten Amendments as ratified by 3/4 of States in 1791)

ARTICLE ONE

Congress shall make no law respecting an establishment of religion, or prohibiting the free exercise thereof; or abridging the freedom of speech, or of the press; or the right of the people peaceably to assemble, and to petition the government for a redress of grievances.

ARTICLE TWO

A well regulated militia, being necessary to the security of a free State, the right of the people to keep and bear arms, shall not be infringed.

ARTICLE THREE

No soldier shall, in time of peace be quartered in any house, without the consent of the owner, nor in time of war, but in a manner to be prescribed by law.

ARTICLE FOUR

The right of the people to be secure in their persons, houses, papers, and effects, against unreasonable searches and seizures, shall not be violated, and no warrants shall issue, but upon probable cause, supported by oath or affirmation, and particularly describing the place to be searched, and the persons or things to be seized.

ARTICLE FIVE

No person shall be held to answer for a capital, or otherwise infamous crime, unless on a presentment or indictment of a grand jury, except in cases arising in the land or naval forces, or in the militia, when in actual service in time of war or public danger; nor shall any person be subject for the same offense to be twice put in jeopardy of life or limb; nor shall be compelled in any criminal case to be a witness against himself, nor be deprived of life, liberty, or property, without due process of law; not shall private property to be taken for public use without just compensation.

ARTICLE SIX

In all criminal prosecutions, the accused shall enjoy the right to a speedy and public trial, by an impartial jury of the State and district wherein the crime shall have been committed, which district shall have been previously ascertained by law, and to be informed of the nature and cause of the accusation; to be confronted with the witnesses against him; to have compulsory process for obtaining witnesses in his favor, and to have the assistance of counsel for his defense.

ARTICLE SEVEN

In suits at common law, where the value in controversy shall exceed twenty dollars, the right of trial by jury shall be preserved, and no fact tried by a jury shall be otherwise reexamined in any court in the United States, than according to the rules of common law.

ARTICLE EIGHT

Excessive bail may not be required, no excessive fines imposed, nor cruel and unusual punishments be inflicted.

ARTICLE NINE

The enumeration in the Constitution of certain rights shall not be construed to deny or disparage others retained by the people.

ARTICLE TEN

The powers not delegated to the United States by the Constitution, nor prohibited by it to the States, are reserved to the States respectively, or to the people.

APPENDIX C

THE TWO GREAT CONVIVENCIAS
ALEXANDRIA, EGYPT

(331 BC/BCE - 1945 AD/CE)

The most detailed and almost breathtaking history of religious tolerance in Alexandria (331 BC/BCE - 641 AD/CE) may be found in books by Theodore Vrettos and by Pollard and Reid (1)

331-30 BC/BCE

The City was founded during 331-332 BC/BCE by Alexander the Great, a student of Aristotle. He was welcomed by Egyptians as pharaoh and a god. As part of a Greek empire, the city became the capital of Egypt. The Greek historian, Diodorus, described it as one of the greatest cities in the world, far surpassing all others in beauty, in the grandeur of its buildings and in riches. While busy conquering the world, Alexander was poisoned and died at age 32 in Babylon (323 BC/BCE). He never saw the city that he planned.

Cleomenes, appointed by Alexander as collector of revenues, was so corrupt that he was deposed and put to death by the Greek general Ptolemy I (Soter). Soter was the first Macedonian governor of Egypt, and died in 290 BC/BCE.

The Ptolemies had become the equivalent of Greek pharaohs. They were military colonists, governing by an immense bureaucracy and amassed great wealth. They "lavished" this on Alexandria (275-215 BC/BCE). In the Near East, Greek became the language of business and law.

Between 285-246 BC/BCE, Ptolemy II enlarged the city of Alexandria. This city had grown so rapidly that by 270 BC/BCE it had to be divided into three districts: the native Egyptian area; the "most thriving" royal Greek-Macedonian area (including other Europeans and Asians); and the Jewish area, almost as large as the Greek. Shortly after Alexander's death, Jews in Alexandria were governed by their own ethnarch (leader) and their own laws. Ptolemy II (Philadelphus) had the Hebrew Scriptures translated into Greek, supposedly employing 70 rabbis.

Alexandrian Jews differed from those in Jerusalem, who were more conservative. In Alexandria wisdom was more important than worship, probably reflecting the influence of the Greek philosophers.

Between 275-215 BC/BCE, Alexandria "was at its zenith, as the center of Hellenistic-Greek culture." The city attracted Greek scholars, Roman emperors, Jewish leaders, Christian bishops (later)and "brilliant minds of the ages." A library larger than any in the known world, magnificent works of architecture, a great university, a harbor and a light-house as ("a wonder of the world") were constructed. The three Punic Wars between Rome and Carthage, in North Africa, raged from 264 to 146 BC/BCE. In the year 200 BC/BCE, Rome was already 500 years old.

Alexandria was about 150 years old. The emperors in Rome continued to be the supreme rulers from 190 BC/BCE until 640 AD/CE, when the city was captured by Arab Muslims. Most of the time, Alexandria was an oasis of religious tolerance, peace and prosperity. Egyptians and Greek worshiped together at the Temple of Serapis, under priests who were followers of the gods, both Zeus and Osiris.

By 59 BC/BCE, Alexandria, the capital of Egypt, had a population of 300,000. It was recognized as the greatest trading city in the known world (2).

30 BC/BCE-300 AD/CE

There are many historical sources regarding the origins and early growth of the *new religion*. Christianity could be dated as beginning with the birth of Christ (c.7-4 BC/BCE) and his crucifixion in 30 AD/CE, ordered by the Roman prefect of Judea, Pontius Pilate, supposedly for having attacked the temple priesthood. At the time, crucifixion of Jews who had revolted against Roman rule had become almost a point of Jewish pride (3)(4)(5)(6).

With the rise of Christianity, there were now conflicts with both the Jews and the pagans, all three insisting they knew the "true" gods.

Perhaps this could be compared with the fabled blind men feeling different parts of an elephant and arguing over the true nature of the animal. In the Middle Ages, duels were fought over how many angels would fit on the head of a pin (7).

In 70 AD/CE, responding to a revolt against Roman rulers, Rome's legions attacked Jerusalem and destroyed the Temple. Jewish dead were estimated at 600,000. Again in 132 AD/CE, in response to an uprising, Jews were driven from most of Judea, which was renamed Syria Palestinia. The number of Jewish dead, including Jewish Christians, was estimated at 850,000. Hadrian was emperor. In Alexandria, the Jews were still relatively safe.

Around 110 AD/CE, a book written by the pagan philo-sopher, Celsus, was such a threat to Christianity that it almost put an end to the new religion. In his book, The True Word, he stated that he "could not understand what madness possessed them (converted Jews) to leave the Law of their fathers and accept a fool who had called himself the Messiah." The Romans and Greeks would not recognize him. His own followers betrayed and deserted him. "As for his resurrection, the same (claim) could be said for many charlatans." Zalmoxis told the Scythians he had come back from the dead. Orpheus, Protesilaus, Heracles and Theseus had all said they had died and risen again. Among other things, he said that, through Moses, God promised the Jews prosperity and earthly dominion. "He bade them to destroy their enemies, sparing neither old nor young." He (Celsus) also quoted Plato as saying that "the

Architect and father of the Universe is not easily found." The "diatribe" continued and is described in considerable detail. However, even Justin, Irenaeus, Polycarp and Ignatius, early defenders of the Jewish Christians, made no effort to respond to Celsus's charges at the time (8).

Finally, Origen (185-254 AD/CE), "renowned Alexandrian teacher, theologian and prolific author," was persuaded to respond with his own book, Against Celsus. For a considerable time after writing his response to Celsus, he was "hounded" by doubts himself. Later many theologians agreed that Origen had really made a most powerful argument against Celsus.

Nevertheless, Origen met with considerable rejection, especially by the Christian Bishop Demetrius in Alexandria, who rejected his doctrines. In 250 AD/CE, while in Tyre, Origen found the anti-Christian persecutions had started again, under the emperor Decius (201-251 AD/CE). During the third century, the emperors were again claiming the status of gods. In Tyre, Origen was arrested and imprisoned. With an iron collar around his neck, he was dragged outside each morning, whipped and his feet crippled. The torture worsened until the emperor (Decius)died in 251 AD/CE. Origen was then released from prison but died in 253 AD/CE.

Alexandrians make a sport of mocking their emperors. This had disastrous consequences with the emperor Caracalla (188-217 AD/CE). He was ridiculed for his small stature and trying to act like another Alexander the Great. He retaliated by ordering the Roman legions to round up every young man and woman in Alexandria, butcher and burn them. His troops obeyed (c. 192 AD/CE).

A general who became Roman emperor, Aurelius Valerius Diocletian, ruled 284-305 AD/CE. He decided to bring order out of chaos and divided the empire into eastern and western halves.

In the East, he would rule and his equal, and fellow "Augustus," Maximium, would rule in the west.

300 – 600 AD/CE

In 303 AD/CE, Diocletian, "to impose order on the empire," decreed the destruction of Christian churches and texts. This stopped in the western Roman empire in 306 AD/CE but continued in the eastern Byzantine Empire until 313 AD/CE. Diocletian died in 305 BC/CE, and Constantine became eastern emperor in 306 AD/CE.

Around 300 AD/CE there were about 3 million Jews in the Empire, mostly in the east in cities, in Alexandria and Carthage. The total number of Christians in the Empire is not known.

After 324 AD/CE, Constantine (c. 285-337 AD), became sole Roman emperor. In 312 AD/CE, he had a vision of the cross prior to a battle where he defeated his rival, Maxentius. It was at that moment that he is said to have converted to Christianity. In 313 AD/CE, with the Edict of Milan, he granted universal religious freedom to pagans, Christians and Jews (9).

After nearly endless Church controversy, Constantine convened the Council of Nicaea (325 AD/CE). To end a destructive conflict, it was in Nicaea that he persuaded the bishops to finally, and by decree, endorse officially that "God and Christ were one." They voted to this as fact.

During the 4th century (300-400 AD/CE), although Roman law authorized Jews to practice their religion, Christian vandal-ism against them was rampant. Many Jewish and Samaritan synagogues were destroyed. The Christian bishops were often referred to as papas (pope, or dear father). Their power in the Mid-east was almost that of a monarch. In Egypt, Osiris, Isis and Apsis were still worshipped.

Ambrose (339-388 AD/CE), bishop of Milan, in 388 AD/CE defended the "righteousness" of burning synagogues and, even murdering the "murderers of Christ (10)."

The pagan emperor Julian, between 361-363 AD/CE, briefly restored paganism. However, by the year 400 paganism was again prohibited, with severe penalties. The people of Alexandria were still considered to be a "free-thinking society." There was continued conflict between these early Christians and non-believers (pagans and Jews), whom they felt might have been tainted by Greek philosophies.

In 395 AD/CE the Roman Empire became permanently divided between Rome in the West and Constantinople in the East.

Constantinople was repeatedly attacked between 400 and 600 AD/CE, by Visigoths, Huns, Ostrogoths, Bulgars and Avars.

Hypatia (370-415 AD/CE), daughter of Theon, renowned mathematician in Alexandria, was a brilliant student at the famous Mouseion University (dedicated in 300 BC/BCE). In the year 400 AD/CE she became the head of Neoplatonist Studies at the University. Students from all the known world competed for her classes. Her "searing intelligence, eloquence and rare beauty" made Alexandria love and glorify her.

Her studies of Plato and Aristotle and, probably, her friendship with the Roman prefect, Orestes, caused the Christian patriarch, Cyril, to fear that she was undermining the Church's authority. Cyril, with his "army of fanatical Nitrian monks," was envious, hated her and spread lies that she was a witch and casting spells. Some citizens, already superstitious, were willing to believe.

Cyril also accused some Jewish leaders of plundering homes of Christians and, at one time had the leaders killed. Cyril's supporters then plundered the synagogues, set fire to Jewish homes and drove some out of Alexandria. When the Roman prefect, Orestes, a Christian, protested against the *marauding clergy,* he was dragged from his carriage, stoned and left unconscious in the street.

Some citizens rushed to his rescue and beat the offending monk to death. Orestes survived and departed the scene with his few soldiers. Reprisal by the "Christian" monks was still ahead.

On a day in 415 AD/CE, during the holy season of Lent, a mob of fanatical monks pulled Hypatia from her carriage, stripped her naked, dragged her to the Christian cathedral, butchered her and burned her body parts.(The first Christian Monks had arrived in Alexandria around 340 AD/CE).

Even more extensive descriptions of the above events can be found in the book by Theodore Vrettos (2001),cited earlier.

Because Constantine the Great (280-337 AD/CE), Byzantine emperor at Constantinople, had decreed that Christianity was compulsory for everyone in the Roman empire, Cyril and his monks felt justified in destroying everything pagan, including the temples of Serapis ad Alexandria, the latter involving the loss of scrolls centuries old. Despite the tensions between cultures, over the centuries, scientists, theologians, writers and philosophers continued to "flow into Alexandria."

In 451 AD/CE at the Council of Chalcedon, the five "great Sees" of the early Church, including Alexandria, were ruled by "patriarchs."

In 529 AD/CE, the emperor Justinian closed the Platonic Academy in Athens and ordered all pagans to become Christians.

Those who refused were exiled and their property confiscated (11).

At Constantinople, the emperor Heraclius (576-642 AD/CE) decided to end the bitter religious disputes that had wracked Egypt since the advent of Christianity. He made the mistake of appointing the patriarch, Cyrus, to carry out his official decree of "monotheism." In the year 631, Cyrus arrived in Alexandria and began persecuting the native Copts, even trying to kill the Coptic patriarch. Heraclius knew nothing of this, but the people of Alexandria were "seething." Cyrus paid little attention to the vulnerability of Alexandria due to the few Greek troops protecting Egypt. As a result, in the year 641, a Persian army easily took Alexandria.

Under Persian rule, Alexandrians had most of their privileges restored. The emperor then suddenly took an interest and routed the Persians. But the city was not free for long.

Rome had ruled Alexandria from 190 BC/BCE until 641 AD/CE. In that year, only about ten years after the death of the prophet, Muhammad, Arab armies took Syria, Palestine and then Egypt. After the "usual looting and raping" in Alexandria, the Arab general, Amru, spared the city's marvelous buildings. He allowed anyone to leave. Those who remained could worship as they wished, but had to pay tribute. Amru's message to the ruling caliph, Omar, in Cairo: "I have taken a city with 4,000 palaces, 4,000 baths, 400 theaters, 1,200 green-grocers and 40,000 Jews who pay tribute."

Eventually, supported by Roman fleets, Alexandria revolted against Arab rule. The Arab general Amru was sent to restore order. A general massacre followed, especially focused on Jews. The great Mouseion Library was totally destroyed, but only after the caliph refused Amru's request to spare it. Many volumes had been destroyed earlier by the fanatical Christian monks who killed Hyptia in 415 AD/CE (12).

After the Arab conquest, Alexandria suffered both physical and spiritual "decay," lasting a thousand years until Napolean's invasion in 1798 AD/CE.

600 – 1700 AD/CE

Although the Christian patriarch, Cyrus, had persecuted members of the native Egyptian Copts around 631 AD/CE, the "ruling monks" now initiated "patriotism" under the banner of religion, wanting to rid Egypt of all foreigners, especially Greeks. Egyptians were again known as *Copts*. Alexandria was *devoured*, with a complete change in language and customs (13).

Arab caliphs ruled Egypt from 641 to 1517 AD/CE. The Turks then took Egypt as part of the Ottoman Empire. The Roman government continued in the Byzantine Eastern Empire until Constantinople fell to the Turks in 1453 AD/CE.

1700-2001 AD/CE

In 1798 AD/CE, Napoleon Bonaparte invaded Egypt, made "striking reforms" in public health and started the science of

Egyptology. However, at the Battle of the Nile, he was defeated by the British fleet under Admiral Horatio Nelson.

In 1805 AD/CE, the Albanian, Mohammad Ali, became pasha of Egypt. The British occupied Egypt in 1882 AD/CE, as "administrators," with Turkish rule continuing. In 1914 Egypt became a British "protectorate" and became an independent state in 1922 AD/CE.

In the year 2001 AD/CE, the Arab Republic of Egypt had an estimated total population of 69.5 million and Alexandria, 3.4 million. Religion was 94 % Muslim and nearly 6% Coptic

Christian. Birth rate: 24.9/1000. (In 1999. The U.S. birth rate: 14.5/1000).

Egypt became a member of the United Nations in 1945 AD/CE.

REFERENCES AND NOTATIONS: ALEXANDRIA, EGYPT

1) Vrettos, Theodor.*Alexandria: City of the Western Mind.* New York and London. The Free Press, 2001. [ALSO SEE: Justin Pollard and Howard Reid: The Rise and Fall of Alexandria – Birthplace of the Modern World. New York. Penguin Books, 2007]

2) Garraty and Gay (Editors).*The Columbia History of the World.* New York. Harper and Row, 1972.[Chapter 18: The Roman Republic][Alexandria pp 185,293,418, 431)

3) Carroll, James. *Constantine's Sword: The Church and the Jews.* New York. Houghton Mifflin Company, 2001. [See Chapter 33 (Convivencia to Reconquista – Cordoba and Alexandria).

4) Armstrong, Karen. *A History of God,*1993.

5) Stearns, Peter (General Editor). *The Encyclopedia of World History. Sixth Edition.* New York. Houghton Mifflin Company, 2001. [See "Early Christianity," pages 91 and 166]

6) Stark,Rodney. *The Rise of Christianity.* San Francisco HarperCollins Publishers, 1996.

7) Alexander,Franz, Slesnick,Sheldon. *History of Psychiatry from Prehistoric Times to the Present,* 1995.

8) Vrettos, Theodore. *Alexandria: City of the Western Mind.* New York and London.The Free Press, 2001.[page 184- 188] 207

9) Carroll, James. *Constantine's Sword: The Church and the Jews.* New York. Houghton Mifflin Company,2001. [See pages 17,180-182]

10) Carroll, James. *Constantine's Sword: The Church and the Jews.*New York. Houghton Mifflin Company, 2001.[See page 201]

11) Garraty and Gay. *Columbia History of the World,* 1972. New York, London. Harper & Row, 1972 See pages 241-243.

12) Vrettos, Theodore. *Alexandria: City of the Western Mind.* New York and London.[The Free Press, 2001. Part Three: The Death of the City, pages 214, 216]

13) Vrettos, Theodore. *Alexandria: City of the Western Mind.* New York and London. The Free Press, 2001.[See pp 207, 212]

CORDOBA, SPAIN

(554-1955 AD/CE)

[REFS 1-19]

In the following confusion of dates, names and wars, perhaps the reader can identify when these "moments in time" occurred. Again, BC/BCE and AD/CE recognizes the different religious calendars.

In addition to Alexandria, Cordoba was another early city, where, for a "moment in time" and "in a kind of paradise," Moors, Jews and Christians lived together in harmony (convivencia), with mutual respect, tolerance, peace and prosperity. This was mostly between 711-1492 AD/CE. In the latter years, Catholics drove out the last Jews and Muslims (1)(2)(3).

Originally inhabited by Celts, Iberians and Basques, Spain became part of the Roman Empire in 206 BC/BCE. In 412 AD/CE, the barbarian Visigoth Atauf, had come south and made himself ruler of Spain.

In 554 AD/CE, the Roman emperor in the east, Justinian, conquered southeast Spain and made Cordoba the provincial capital. In 587 AD/CE, the Iberian peninsula became Catholic, with the conversion of the Visigoth king, Recarred.

After 600 AD/CE, Jews were forced to accept Catholic baptism.

Ironically, perhaps, this "peaceful co-existence" seemed mostly during the later periods of Muslim rule. "Muslim Spain" refers to the periods when various Muslim factions invaded and occcupied the early cities of the Iberian peninsula.

Meanwhile, in far off Mecca, around 570 AD/CE, Muhammad ibn Abdullah, the founding "Prophet" of Islam, was born. He died in 632 AD/CE. This was to be the most significant religious event since the birth of Jesus (c.4 BC/BCE-7 AD/CE) (4).

600-800 AD/CE

In 638 AD/CE the Muslims conquered Jerusalem, setting the stage for the later "Christian" crusades to retake the "Holy Land." By 661 AD/CE, the Arab-Muslim Empire included Arabia, Syria, Armenia, east almost to India, and by 670 AD/CE, the north African coast as far west as Morocco.

The first Muslim invasion of Spain was between 711-715 AD/CE, by Arabs and Berber tribesman from Morocco. An Arab culture in Spain would last for 800 years. This was the beginning of prolonged war between Muslims and Catholic Christians. The Jews had welcomed the Muslims (5).

The Muslim foray continued into France but was halted by Charles Martel at the Battle of Tours (732 AD/CE). Muslims retreated to Spain, where the emirs (governors) ruled most of Spain until 756 AD/CE. In Iberia, Christians were hostile to these Muslims, but Jews were receptive. Some Jews became quite wealthy through commerce, textile manufacture and in the state-supported slave trade.

In 756 AD/CE, an Umayyad (Sunni) prince, Abd al-Rahman I, invaded Spain, defeated the ruling emir (governor) of Cordoba and established his own Islamic state, independent from the Abbasid caliph in Damascus.

Abd al-Rahman I had escaped in 750 AD/CE, in Damascus, from the slaughter of most of his kinsmen when the Abbasids revolted against the ruling Umayyad caliphs. This new "Umayyad dynasty of Cordoba," mostly under the al-Rahmans, lasted until 1031 AD/CE, with the last Umayyad caliph, Hisham III. To even begin to understand the centuries-old rivalry between the Sunni/Umayyad and Shiite, one is referred to the following texts (6)(7).

After 756 AD/CE, when Abd al-Rahman I initiated the "Umayyad dynasty of Cordoba," the Jews and Christians were protected because they were "People of the Book," as were Muslims. They were treated well but forced to pay a special tax. Soon, Muslim "dissidents," were opposing al-Rahman, and were being supported by the Christian Emperor Charlemagne.

In 777 AD/CE, this Christian king of the Franks, invaded Spain but, by 810 AD/CE, had conquered only parts of northeastern Spain, such as Barcelona. (In 800 AD/CE, he was crowned the first Holy Roman Emperor.)

800-900 AD/CE

In 837 AD/CE, a revolt by Christians and Jews in Toledo was suppressed by al-Rahman II, but Christian "fanatics" continued to be active, especially in Cordoba. During the reign of al-Rahman II (822-853 AD/CE), another Christian, Alfonso II of Leon invaded Aragon but was defeated.

Muhammad I (852-853 AD/CE) put down another Christian uprising in Cordoba and waged war against the Christian cities of Leon, Galicia, and Navarre.

A vast Islamic Empire reached the "height of (Islamic) civilization" between 750-1258 AD/CE. It was established mainly by the swords of the majority, militant Sunni-Umayyad faction of Islam (8)(9).

900-1000 AD/CE

Abd al-Rahman III (912-961 AD/CE), considered to be the most gifted of the Spanish Umayyads, became caliph in 929 AD/CE. He placed himself even above the caliphs of Baghdad. Spain was at peace and began to thrive. This was the so-called "green revolution" in Spain. Libraries and universities sprang up alongside palaces and gardens. Under the rule of this caliph, Christians held their services in the Great Mosque of Cordoba. (Today a Catholic cathedral stands in the place of the mosque.)

During this 10th century in Cordoba, the Umayyad caliph Hakim II, also known as al-Mustansir (ruled 936-976 AD/CE), tried to rival the Abbasid caliph in Baghdad, importing scientific and philosophical books, making the Cordoban university and library similar to those in Baghdad. Jews were taught Arabic by Muslim scholars, mastered the Koran and the philosophies of Plato and Aristotle. Jews and Muslims read both the Koran and Hebrew scriptures.

Around 950 AD/CE, Muslim Cordoba had a population of some 500,000 people, 1,600 mosques, 900 public baths, 80,455 shops and a library of 400,000 volumes. It was described as "the ornament of the world" by the Saxon nun, Roshwitha.The old aristocratic Visigoths were replaced by a rich middle-class of Christians and Jews. Muslim wars continued with the Christian cities of Leon and Navarre (10).

In Cordoba, three outstanding Muslim scholars were Ibn Masarrah (died 931), al-Majriti (died 1009) and al-Kirmani (died 1086). Abu al-Qasim al-Zahrawi produced a medial encyclopedia between 936-1013 AD/CE. In the late 10th century, under the rule of the Umayyad caliph, Abu Amir al-Mansur (Almanzor), this near-paradise began to change. In the late 900s, al-Mansur plundered churches. Any Muslim who converted to Christianity was immediately put to death. (He did even worse things.)

1000-1100 AD/CE

From about 1016 AD/CE, following a long "civil war" between the Umayyads and Berbers, Muslim Spain was reduced to a "score of petty kingdoms, making a Christian reconquest easier."

Finally is 1031 AD/CE, leading families in Cordoba decided to abolish the Umayyad caliph completely. Hisham III was the last

Umayyad caliph. This brought about the era of "petty kings" of competing city-states, the "Muluk al-Tawa'if dynasties" (1020-1086 AD/CE). By the year 1086 these city-state s were "mostly absorbed" by the *Abbasids*.

This also became the "golden age" of Jewish culture, where each petty monarch welcomed their talents and trustworthiness and their elegance in Arabic speech and manners. An example was the "spectacular career" of the multi-talented Samuel ibn Nagrela from the Academy of Cordoba, who quickly became the vizier of Granada and a byword to Jew and Muslim alike. In 1067 AD/CE, a "resentful Muslim mob" killed ibn Nagrela's son and massacred hundreds of Jews in Granada (11).

In 1085 AD/CE, Alfonso VI of Castile started the Christian reconquest (reconquista) of Spain, but only succeeded in capturing Toledo.

Needing assistance in defending themselves against Alfonso, the Abbasids invited "puritanical" Berber Almoravid leader, Yusuf ibn Tashfin, to Spain to assist in defense against Alfonso. These Muslims, acting together, defeated Alfonso in 1086 and annexed "Moorish" Spain: Cordoba, Toledo, Medina, Sidonia, Seville, Merida, and Saragossa. They drove any remaining Roman Catholics to the North and West.

This peaceful co-existence (convivencia) began to break apart when the Christian "crusading fervor" began in Europe in 1096 AD/CE. In 1096 AD/CE, "Christian" Crusaders massacred Jews — "killers of Christ" — at Mainz.

Abu Bakr ibn Yahya ibn Sayigh (c.1095-1139 AD/CE), philosopher and physician, wrote about Aristotle and Plato. He was considered to be the most eminent Muslim scholar of the time (12).

1100-1200 AD/CE

After the death of Yusuf in 1106 AD/CE, this Moorish empire quickly disintegrated. Alfonso VI resumed the Christian Reconquest (reconquista) of Spain. In 1146 AD/CE, Alfonso VII, did succeed in recovering Cordoba (13).

In 1135 AD/CE, in Cordoba, Alfonso VII was crowned as a Christian "emperor" and "felt he was the equal" of any Holy Roman Emperor of the time.(Note: The first Holy Roman Emperor was Charles, grandson of Charles Martel, known as Charlemagne, who was crowned in 800 AD/CE by Pope Leo III.)

In 1145 AD/CE, a Muslim "Almohade avalanche" had swept across North Africa to Morocco. Between 1147-1276 AD/CE, the persecution of Jews in North Africa was "unmatched in (its) brutality by any Christian state prior to the anti-Semitic excesses of 19th century czarist Russia (14). "

The second major invasion of Spain from Morocco (1145 AD/CE) by "militant Muslims," was by "puritanical" *Almohads.* They were believed to be resentful of the intermingling between Muslims and infidels—and possibly envious of their "soft lives." For nearly 20 years the Almohades fought the Iberian caliphs, Christians and Jews before gaining control of south and central Spain. Many Muslims, Christians and Jews died. Most Jews fled to Egypt or Christian Spain (15).

By 1146 AD/CE, Alfonso VII had recaptured Cordoba, placed a cross atop the Great Mosque, where Catholic and Muslims had worshipped together - perhaps even from when it was built between 700-800 AD/CE.)

However, the Almohad Muslims soon recaptured the city, removed the cross and became "ruthless" toward Jews and Christians. By 1159 AD/CE, Jews were being forcibly converted to Islam or murdered.

The height of Muslim learning was reached by ibn-Rushd (Averroes)(c.1126-1198 AD/CE, philosopher, physician, commentator on Plato and Aristotle, teacher of Christians (16).

Moses ben Maimon (1135-1204 AD/CE), a native of Cordoba, the most famous of Jewish sages, wrote in Arabic, not Hebrew, regarding the great philosophers and scientists, including Plato, Aristotle, Euclid, Ptolemy, Pythagoras and almost all the Muslim philosophers. Jews taught Christians, especially in Castile and Catalonia. For the time, ethnic and religious creeds meant less than one's region, social role and work.

In 1159 AD/CE, the ruthlessness of the Almohades caused many Jews to flee. Other converted to Islam or were killed. The brilliant 24 year old Maimonides fled to Egypt, where he codified all Jewish law, synthesizing the Jewish faith with Aristotelian philosophy.

1200-1492 AD/CE

The fourth (Christian-Catholic) Crusade occurred between 1200-1204 AD/CE. As special Crusade between 1208-1213 AD/CE was made against "heretical" Albigensians (Catharists of Albi) in Southern France, by Innocent III. Many died.

In 1212 AD/CE, the Almohades were finally and permanently defeated by Christian kings, Peter II of Aragon and Alfonso VIII of Castile. They were driven from Spain and only "local" Muslims dynasties remained. A cross was restored atop the Great Mosque. Many Muslims retreated to Granada or North Africa and some became Catholic Christians.

At this time, Iberia had a population of about 3 million with several hundred thousand Jews. Aware of the ominous beginnings of the Catholic Inquisitions, the Jews braced themselves. (This seemed wise in view of Ferdinand and Isabella later expelling them from Spain in 1492 AD/CE.)

In 1231 AD/CE, Pope Gregory IX had started the *Inquis-itions* with his *Excommunicamus*, establishing the Dominican and Franciscan courts. The Spanish Inquisition began in 1492 AD/CE, when the Grand Inquisitor, Torqemada, forced Jews to convert or be expelled. The Spanish Inquisition ended in 1543 AD/CE (17).

A "tug-of-war" between Christians and Muslims continued until Christian king Ferdinand III (1217-1252 AD/CE) retook Cordoba in 1236 AD/CE and Seville in 1248 AD/CE. He called himself the "king of three religions." The Moors were not disliked and intermarriages took place during the 13th century (18)(19).

By 1250 AD/CE, most of Iberia was ruled by the Christian kings of Aragon, Castile and Portugal, with only the Arab emirate of Granada surviving.

In 1492 AD/CE, Roman Catholicism was the official religion and Jews were expelled. The last Muslims were forced to leave in 1505 AD/CE.

In 1991 AD/CE, the Kingdom of Spain had a population estimated at 40 million. The literacy rate was 95 percent.

Religion was 99 percent Roman Catholic. Birth rate was 9.3 per 1,000 population. (Note: the U.S. birthrate in 1999 was 6.1 per 1000.) Spain became a member of the United Nations in 1955 AD/CE.

REFERENCES AND NOTATIONS: CORDOBA, SPAIN

1) Carroll,James. *Constantine's Sword .Boston and New York.* Chapter 33:*Convivencia and Reconquesta,* pages 322-342.
2) Stearns,Peter (General Editor).*The Encyclopedia of World History.*Boston and New York. Houghton Mifflin,2001. [See Muslim Spain, pages 179 and 218-221]
3) Espinosa,John(Editor).*The Oxford History of Islam.* Oxford University Press, 1999. [See page 33 and *Chronology,* pages 691-696 (c. 570-1998 AD/CE)]
4) Daniels,Patricia;Hyslop,Stephen. *Almanac of World History.* Washington,DC. National Geographic,2003.[See *Muhammad and Islam (570-1000)*[NOTATION: This 384 page book includes many maps and is beautifully illustrated. Almost encyclopedic, it starts with Human Evolution and prehistory (3000 BC)- and ends with early 2001 AD/CE.
5) Garraty and Gay (Editors). *Columbia History of the World,* 1972.{ See pages 270, 294]
6) Esposito,John (Editor).*The Oxford History of Islam,*1999. [See Chapter One to even begin to clarify the tangled division between Sunni-Shiite]
7) Armstrong,Karen. *Islam.* New York. The Modern Library, 2000.
8) The World Almanac- 2005. New York. St. Martin's Press, 2005.[See page 508]
9) Lewis, Brenda (General Editor). *Great Civilizations.* Bath, UK. Paragon Publishing 2002. [Page 146]
10) Stearns,Peter. *The Encyclopedia of World History- Sixth Edition.* New York. Houghton Mifflin Company, 2001. [Pages 179, 219]
11) Garraty and Gay. *Columbia History of the World,* 1972. New York, London. Harper & Row, 1972.[Pages 294-295]
12) Esposito, John. (Editor). *The Oxford History of Islam,* 1999.[Pages 284-285]
13) Stearns, Peter (General Editor). *The Encyclopedia of World History. Sixth Edition.* New York. Houghton Mifflin Company, 2001.[See pages 218-222]
14) Garraty and Gay. *Columbia History of the World,* 1972. New York, London. Harper & Row,1972.[See page 289]
15) Stearns, Peter (General Editor). *The Encyclopedia of World History. Sixth Edition.* New York. Houghton Mifflin Company, 2001.[See pages 218-210]
16) Carroll, James. *Constantine's Sword: The Church and the Jews.* New York. Houghton Mifflin Company, 2001. [Chapter 33: Convivencia and Reconquista, pages 322- 342]

17) Carroll, James. *Constantine's Sword: The Church and the Jews.* New York. Houghton Mifflin Company, 2001. [See page 262]
18) Stearns, Peter (General Editor). *The Encyclopedia of World History. Sixth Edition.* New York. Houghton Mifflin Company, 2001. [See pages 219-221]
19) Carroll, James. *Constantine's Sword: The Church and the Jews.* New York. Houghton Mifflin Company, 2001. [See Chapter 33]

APPENDIX D

QUOTATIONS, FAMOUS AND INFAMOUS

[ALL 48 REFERENCES ARE SHOWN WITHIN THE TEXT]
[Note: BCE, Before Current Era, and CE, Current Era, are used in place of BC or AD, recognizing other religious calendars]

SUN TZU, fabled Chinese general (circa 600-500 BCE), has been quoted:

"In martial arts, it is important that strategy be unfathomable, that form be concealed, and that movements be unexpected, so that preparedness against them be impossible."

Pretend friends are enemies, and enemies are friends. Have spies everywhere.

[Cleary, Thomas (translator). *The Art of War: Sun Tzu.* Boston. Shambhala, 1988.]

Some believe this treatise was actually written by Lorenzo Ricci, 18th Superior General of the Society of Jesus (1758-1775 CE, who has sometimes been referred to as the *Black Pope.* No current history book on ancient China mentioned *Sun*-tzu.

[Saussy, F. Tupper. *Rulers of Evil.* New York. HarperCollins Publishers, 1999. See pages 85-95]

[Garraty and Gay. *Columbia History of the World,* 1972. See *Early China* pages 107-135]

[Stearns, Peter. *The Encyclopedia of World History,2001.* See multiple references to China]

Sun-tzu is also quoted as saying:

"The worst policy is to attack cities."

In this same editorial, cited below, tragic examples are given, including attacks on Stalingrad, Manila, Grozny, and Mogadishu (Somalia). Massive civilian and military casualties resulted. The following reference was prompted by the "coalition forces" invasion of Iraq in April 2003.

[See Guardian Newspapers 04/03/03,at www.buzzle.com/editorials]

For more detailed information regarding Sun Tzu and China between 722 and 470 BC/BCE, you may log on the Internet at: www.clearbridge.com/sun_tzu_history.htm or www.clearbridge.com/china_timeline.htm

Compare:

> *"The expert in using the military subdues the walled cities without launching an attack, and crushes the enemy's state without protracted war."*

[Carr, Caleb. *The Lessons of Terror: A History of Warfare Against Civilians: Why It Has Always Failed and Why It Will Fail Again.* New York. Random House, 2002.[See page 103]

MARCUS CICERO (106-43 BC/BCE),Roman philosopher, barister, senator, champion of the law, was murdered along with some 130 senators and 2,000 equestrians declared "public enemies" by the second Triumvirate (Antony, Lepidus and Octavian.) They took power after the murder of Julius Caesar in 44 BC/BCE. Cicero is quoted:

> *"A nation can survive its fools, and even the ambitious, but it cannot survive treason from within. An enemy at the gates is less formidable, for he is known and he carries his banners openly. But the traitor moves among those within the gates freely ... for the traitor appears not a traitor — he speaks in the accents familiar to his victim, he wears their face and garments, and he appeals to the baseness that lies deep in the hearts of all men. He rots the soul of a nation —he works secretly and unknown in the night to undermine the pillars of*

a city — he infects the body politic so that it can no longer resist ... A murderer is less to be feared."

[Garraty and Gay. *Columbia History of the World,* 1972. New York, London. Harper & Row,1972. See the *Roman Republic,* pages 199-200]

[Veon,Joan. *The United Nations' Global Straightjacket.* Oklahoma City.Heartstone publishing, 2000.See page 199]

"Barbarian kings schemed against one another for the favor of influential Romans, sending gifts to senators, and entertaining generals with the obsequiousness of humble subjects, who nevertheless ventured to present their Roman majesties with jewelry, money, tables, chairs and bedsteads of gold, and beautiful slaves. [Cicero].

Note the looting of Rome's colonies and provinces and also compare Cecil Rhodes's quotation in 1897 AD/CE).

[Petersson, Torsten. *Cicero: A Biography.* Berkeley, CA. University of California Press, 1920. See page 42]

TITUS LUCRETIUS (98-55 BC/BCE), Roman poet, author of the *Nature of Things (De rerum natura),* following Democritus's theory of a world atoms:

"There is no hell except here." This statement was in response to the "rising cult of heaven and hell among the people of Rome."

[Durant, Will. *The Story of Philosophy.* New York. Simon and Schuster, 1926.]

Compare George Santayana, who agreed with Lucretius, that "fear first made the gods ."

[Santayana, George. *The Life of Reason.* New York. Prometheus Books, 1998. See page 192]

%

POPE BONIFACE VIII, in 1302 A.D., contesting the authority of King Philip IV of France, reportedly stated:

> *"It is absolutely necessary for the salvation of every human creature to be subject to the Roman Pontiff."*

[Saussy, F. Tupper. *Rulers of Evil.* New York. HarperCollins Publishers, 1999.See page 37]

%

BARTHOLOME DE LAS CASAS, a young priest who owned a Cuban plantation with Indian slaves, around 1508 AD/CE, wrote as an eye-witness in his *History of the Indians:*

> *"Endless testimonies ... prove the mild and pacific temperaments of the natives ..." "But our work was to exasperate, ravage, kill, mangle and destroy; small wonder, then if they tried to kill one of us now and then." "The admiral (Columbus), it is true, was blind as those who came after him, and he was so anxious to please the King (of Spain) that he committed irreparable crimes against the Indians ."*

[Zinn, Howard. *A People's History of the United States: 1492-Present.* New York. Harper Perennial, 1995. See pages 5-7]

%

WILLIAM BRADFORD, Governor of the English Pilgrim's colony at Plymouth, around 1621 AD/CE, "lashed out at the communal system" fostered by the London merchants who financed the venture:

> *"The vanity and conceit of Plato and other ancients...that the taking away of property, and bringing (it) in community ... would make them happy and flourishing ; as if they were wiser than their God...(However it) was found to breed much confusion and discontent, and retard much employment that would have been to their benefit and comfort."*

[Still, William. *New World Order: The Ancient Plan of Secret Societies*. Lafayette, LA. Huntington House Publishers, 1990.See page 58]

%%%

VOLTAIRE, pen-name of Francois Marie Arouet (1694-1778 AD/CE):

> *"Those who can make you believe absurdities can make you commit atrocities."*

%%%

JOHANN WOLFGANG VON GOETHE, (1794-1832 AD/CE), speaking through his character, Faust, in Part One of his eponymous drama:

> *"I have, alas! Philosophy, Medicine, Jurisprudence too, And to my cost Theology, With ardent labour, studied through. And here I stand, with all my lore, Poor Fool, no wiser than before."*

%%%

THE FRENCH PARLIAMENT, on August 6, 1762 AD/CE, passed a resolution as follows:

> *"The Society of Jesus (Jesuits) by its very nature is inadmissible in any properly ordered State as contrary to natural laws, attacking all temporal and spiritual authority, and tending to introduce into Church and State, under the specious veil of a religious Institute, not an Order truly aspiring towards evangelical perfection, but rather political organization whose essence consists in a continual activity, by all sorts of ways, direct and indirect, secret and public, to gain absolute independence and then the usurpation of all authority...They outrage the laws of nature and as enemies of the laws of France should be irrevocably expelled ".*

[Saussy, F. Tupper. *Rulers of Evil*. New York. HarperCollins Publishers, 1999.See page 107]

[Paris, Edmond. *The Secret History of the Jesuits: Translated from the French -1975*. Ontario, CA. Chick Publications. See pages 82-87]

%

THOMAS JEFFERSON, 3rd U.S. President (1802-1808), in 1776 is quoted in regard to a threat to American's Rights of Life, Liberty, and the Pursuit of Happiness:

> *"That whenever any Form of Government becomes destructive of these ends, it is the Right of the People to alter or abolish it, and to institute new Government, laying its foundation on such principles likely to effect their Safety and Happiness."*

www.bilderberg.org/roundtable/Title50.htm.

[Also see: *Declaration of Independence* by Thomas Jefferson, July 4,1776]

Jefferson is also quoted:

> *"If the American people ever allow the banks to control the issuance of their currency, first by inflation and then by deflation, the banks and corporations that will grow up around them will deprive the people of all property until their children will wake up homeless on the continent their fathers occupied. The issuing power of money should be taken from the banks and restored to Congress and the people to whom it belongs. I sincerely believe the banking institutions are more dangerous to liberty than standing armies."*

[Sutton, Antony. *The Federal Reserve Conspiracy*, 1995. See page 6]

%

ABRAHAM LINCOLN, U.S. President, on November 21, 1864, wrote/said:

> *"I see in the near future a crisis approaching that unnerves me and causes me to tremble for the sake of my country...*

Corporations have been enthroned and an area of corruption in high places will follow, and the money power of the country will endeavor to prolong its reign by working upon the prejudices of the people until all wealth is aggregated in a few hands and the Republic is destroyed."

[Caldicott,Helen. *The New Nuclear Danger: George W. Bush's Military-Industrial Complex*. New York. The New Press, 2002. See Chapter Four: Corporate Madness and the Death Merchants]

A better known quotation

"If once you forfeit the confidence of your fellow citizens, you can never regain their respect and esteem. It is true that you may fool all the people some of the time. You can even fool some of the people all the time. But you can't fool all the people all of the time."

%

BENJAMIN DISRAELI, British Prime Minister, around 1886 AD/CE, told the British House of Commons,

"It is useless to deny, because it is impossible to conceal, that a great part of Europe, the whole of Italy and France and a great portion of (then fragmented) Germany — to say nothing of other countries — is covered with a network of secret societies...
And what are their objects? They do not attempt to conceal them. They do not want a constitutional government ...they want to change the tenure of land, to drive out the present owners of soil and put an end to ecclesiastical establishment (churches)."

[Marrs, Jim. *Rule of Secrecy*.HarperCollins Publishers, 2000. Quoted on page 13]

Disraeli is also quoted as having a character in his book say:

"So you see ...the world is governed by very different personages from what is imagined by those who are not behind the scenes."

(Disraeli was close to the international banking *House of Rothschild* and may have had Lionel Rothschild in mind when creating the character Sidonia .)

[Ferguson, Niall. *The House of Rothschild: Money's Prophets,* 1798-1848. New York. Penguin Books, 1999.]

[Arendt, Hannah. *The Origins of Totalitarianism.* NY,NY Harvest Books/Harcourt, Inc, 1966 and 1974. Pp.68-78]

[Still, William. *New World Order.* Lafayette, LA. Huntington House Publishers, 1990.See Chapter 12: Central Banking, the Council on Foreign Relations and FDR]

%

SIR JOHN ACTON, British House of Lords, Regius Professor of History at Cambridge, in 1887:

"Power tends to corrupt, and absolute power corrupts absolutely."

Although he was Catholic, he was allegedly referring to the pope .

[Cited in *the Oxford English Dictionary* of Quotations – Third Edition. Oxford University Press, 1980.]

[*Carrol,James. Constantine's Sword.* Boston, New York. Houghton Mifflin, 2001. See page 573]

%

VLADIMIR ILYICH ULYANOV-LENIN (1870-1924), reportedly said that the establishment of a central bank was ninety percent of communizing a country.

[Still, William. *New World Order: The Ancient Plan of Secret Societies.* Lafayette, LA. Huntington House Publishers, 1990.]

%

THEODORE ROOSEVELT , U.S. President (1901-1909), in 1893, in an address to the Naval War College, had said:

"All the great masterful races have been fighting races ... No triumph of peace is quite so great as the supreme triumph of war."

In 1897, just prior to the war with Spain, he wrote to a friend:

"In strict confidence ...I should welcome, almost any war, for I think this country needs one."

[Zinn,Howard. *A Peoples History of the United States: 1492-Present.* New York. Harper Perennial, 1995. Pages 290, 293]

Yet in 1905, he (TR) won the Nobel Peace Prize for mediating an end to the Russo-Japanese War .

[Kunhardt, Philip, Jr., Philip III and Peter K. *The American President.* New York. Riverhead Books/ Penguin Putnam, 1999]

Prior to Roosevelt's presidential campaign, two of "J.P. Morgan's men," including the chairman of U.S. Steel, had "arranged a general understanding with Roosevelt by which ... they would cooperate in any investigation by the Bureau of Corporations in return for a guarantee of their companies' legality." (And yet, he was known as a *trust-buster.*)

[Wiebe,Robert. *The Search for Order, 1877-1920.* New York. Hill & Wang, 1966.Page 342]

///

CECIL RHODES, founder of Rhodesia (now called Zimbabwe) was quoted in *The Economist*, March 27, 1993, as saying (circa 1897):

"We must find new lands from which we can easily obtain raw materials and at the same time exploit the cheap slave labour

that is available from the natives of the colonies. The colonies would also provide a dumping ground for the surplus good produced in our factories."

This philosophy seems to have typified all that has followed in the exploitation of *undeveloped and developing countries in the "third world"*, which continues today.

///

GEORGE SANTAYANA (1863-1952), Harvard professor of philosophy, is quoted in 1905 –

"Those who cannot remember history are doomed to repeat it."

[*Life of Reason, Reason in Common Sense*. New York. Scribners,1905. Page 284]

///

WOODROW WILSON, U.S. President (1913-1921), is quoted in 1907 at a Columbia University lecture:

"Concessions obtained by financiers must be guarded by ministers of state, even if the sovereignty of unwilling nations be outraged in the process ... the doors of the nations must be battered down."

Yet in 1911, while governor of New Jersey, he is quoted as saying:

"The greatest monopoly in this country is the money monopoly. So long as that exists, our old variety of freedom and individual energy of development are out of the question."

In 1913 he signed the Federal Reserve Act, rather than veto it. In 1914, he said he supported "the righteous conquest of foreign markets."

After Congress approved this privately owned central bank, the *Federal Reserve System,* Wilson, in his book, *the New Freedoms* stated:

> *"Some of the biggest men in the United States, in the field of commerce and manufacture, are afraid of something. They know there is a power somewhere so organized, so subtle, so watchful, so interlocked, so complete, so pervasive, that they had better not speak above their breath in condemnation of it".*

Woodrow Wilson tried to keep the U.S. neutral early in World War One, but on April6, 1917, the U.S. finally declared war on Germany. The Treaty of Versailles in January 1919 was followed by the *Covenant of the League of Nations,* signed in June and July. Despite all efforts by Wilson, the isolationist U.S. Congress refused to join the League. It seems significant that the Carnegie Board of Directors (steel) had urged Wilson to see that the war not end "too quickly." Bernard Baruch, head of the War Industries Board, and John D. Rockefeller (Standard Oil), allegedly "reaped profits from the war of some $ 200 million."

[Still,William. *New World Order,* 1990.]

//,

JOHN HYLAN, former New York mayor, in 1922, stated:

> *"The real menace to our Republic is the invisible government which, like the giant octopus, sprawls its slimy length over our city, sate and nation . . . . At the head of this octopus are the Rockefeller Standard Oil interests and a small group of powerful banking houses, generally referred to as the international bankers,(who) virtually run the U.S. government for their own selfish purposes."*

[Marrs,Jim. *Rule by Secrecy.* New York. HarperCollins, 2000].

%

REGINALD MCKENNA, former chancellor of the Exchequer of England (1915-1916), and chairman of the Board of the Midlands Bank of England, in January 1924 told his stockholders:

> *"I am afraid the ordinary citizen will not like to be told that the banks can, and do, create money ... and they who control redit of the nation direct the policy of Governments and hold in the hollow of their hands the destiny of the people)."*

[Perloff,James. *The Shadows of Power: The Council on Foreign Relations and the American Decline.* Appleton, WI. Western Islands, 1988. See page 20]

[Still,William. *New World Order.* Lafayette, LA. Huntington House Publishers, 1990]

[Garraty and Gay. *The Columbia History of the World*, 1972]

%

FRANKLIN D. ROOSEVELT, U.S. President, in a letter dated November 23, 1933, to Woodrow Wilson's chief adviser, Colonel Edward House:

> *"The real truth of the matter is, as you and I know, that a financial element in the large centers has owned the government ever since the days of Andrew Jackson."*

In 1935, FDR added to the Great Seal of the United States, the "all-seeing eye," from ancient Egypt, with the underlying Latin phrase "Novus Ordo Seclorum", or New Order of the Ages. (Also note the back of a U.S. one dollar bill,the all-seeing eye over a pyramid with the same "Novus Ordo Seclorum". The other Great Seal motto, "Annuit Coeptis," allegedly relates to ancient Greek mythology and the god Jupiter. Supposedly, it roughly translates as "God hath favored this undertaking." This Latin inscription seems to suggest a connection between the Roman Church's Pontifex

Maximus and our "founding fathers." The "all-seeing eye" of Osiris has been a symbol of Masonry)

[Still,William. *New World Order*,1990.See pages 24,65]

%

FELIX FRANKFURTER, U.S. Supreme Court Justice (1939-1962) was quoted:

"The real rulers in Washington are invisible, and exercise power from behind the scenes."

[Marrs, Jim. *Rule by Secrecy*,2000. See page 3]

%

HARRY SHIPPE TRUMAN, a U.S. Senator, on March 27,1942, as Chairman of the Senate Internal Security Committee, in regard to the Rockefeller's Standard Oil of New Jersey was quoted:

"Even after we were in the war, Standard Oil of New jersey was putting forth every effort of which it was capable to protect the control of the German government over vital war material. As Patrick Henry said: 'If this be treason – and it certainly is treason – then make the most of it." (Compare Cicero, as quoted in 50 BC/BCE).

%

HANNAH ARENDT (1906-1976), political theoretician and author, writing on imperialism:

"The decade immediately before the imperialist era (of empire building), the seventies of the last century (1870s), witnessed an unparalleled increase in swindles, financial scandals, and gambling in the stock market."

And writing on secrecy:

"The totalitarian movements have been called 'secret societies established in broad daylight'."

[Arendt, Hannah. *The Origins of Totalitarianism.* Harcourt, 1951, 1966, 1968, and 1994]

%

JAMES WARBURG, banker, son of Council of Foreign Relations (CFR) founder Paul Warburg, on February 17, 1950, told the Senate:

"We shall have world government whether or not we like it. The only question is whether world government will be achieved by conquest or consent."

[Still, William. *New World Order.* Lafayette, LA. Huntington House Publishers, 1990. See page 174]

%

DWIGHT D. EISENHOWER, U.S. President, has been quoted as saying on April 16, 1953:

"Every gun that is made, every warship that is launched, every rocket fired signifies, in the first sense, a theft from those who hunger and are not fed, those who are cold and are not clothed. This world in arms is not spending money alone, it is suspending the sweat of its laborers, the genius of its scientists, the hopes of its Children — this is not a way of life at all in any true sense. Under the cloud of war it is humanity hanging on the cross of iron."

In his farewell address, January 17, 1961:

"In the councils of Government, we must guard against the acquisition of unwarranted influences, whether sought or unsought, by the military-industrial complex. The potential for the disastrous rise of misplaced power exists and will persist. We must never let the weight of this combination endanger our liberties or democratic processes."

[Borkin, Joseph. *The Crime and Punishment of I.G. Farben: The*

Unholy Alliance Between Hitler and the Great Chemical Combine. New York. Barnes and Noble, 1978]

///,

WILLIAM JENNER, U.S. Senator, on February 23, 1954, is quoted:

"Today the path to total dictatorship in the United States can be laid by strictly legal means, unseen and unheard by the Congress, the President or the people . . .Outwardly we have a constitutional government. We have operating within our government and political system, another body representing another form of government, a bureaucratic elite which believes our Constitution is outmoded and is sure it is on the winning side … All the strange developments in foreign policy agreements may be traced to this group … This political action group has its own political support organizations, its own pressure groups, its own vested interests, its foothold within our government, and its own propaganda apparatus."

[Perloff,James. *The Shadows of Power: The Council of Foreign Relations and the American Decline.* Appleton, WI. Western Islands Publishers, 1988]

Carroll Quigley, professor of history at Georgetown, Princeton, and Harvard in 1966 is quoted:

"The history of the last century shows … that the advice given to governments by bankers ... was consistently good for bankers, but was often disastrous for governments, businessmen, and the people generally. Such advice could be enforced, if necessary, by manipulation of exchanges, gold flows, discount rates, and even levels of business activity".

[Quigley, Carroll. *Tragedy and Hope.* New York. Macmillan, 1965. Page 62]

///,

EDMUND PARIS, author of "The Secret History of the Jesuits," translated from the French in 1975, states:

"No one can be deceived — and the Jesuits less than others. A general disarmament would toll the knell of the Roman Church as a world power."

[Paris,Edmund.Ontario, CA. Chick Publications,1975]

%

ZBIGNIEW BZREZINSKI, in 1985, at the Gorbachev State of the World Forum:

"The new world order (will) come step by step and stone by stone."

[Veon,Joan. *The United Nations Global Straightjacket*, 2000. Page 320]

%

William Still, in his book *New World Order* (1990) states:

"When you control the credit of a nation, you control its economy."

He also states:

"Fortunately there are very few completely evil men in the world. The vast majority of members in these (secret) groups are merely deceived. Their minds can be changed when presented with the truth. It's important to remember that the people best able of helping defeat the machinations of the secret societies are its members."

(See page 182).

"The secret architects of this 'Great Plan' (New World Order) are not benign humanitarians, as they would have us believe, but are men in the service of evil. Their 'government of nations' is a deception, hiding, in reality, an iron-clad, world dictatorship."

(See page 6).

"Therefore all this banter about political 'rights' and 'lefts' really doesn't make much sense. After all rightists can be dictators. Leftists can be dictators. That is not the point. Once a condition of chaos is created in a government, then freedom inevitably suffers."

(Page 18) (Compare provisions of the 2001 *U.S.A. Patriot Act* on the human rights guaranteed by the U.S. Constitution.)

DAVID ROCKEFELLER, of the Chase Manhattan Bank, is quoted as saying at the Bilderberg meeting in Baden, Germany, in June 1991:

"It would have been impossible for us to develop our plan for the world if we had been subjected to the lights of publicity during those years. But, the world is more sophisticated and prepared to march towards a world government. The supranational sovereignty of an intellectual elite and world bankers is surely preferable to the national auto-determination practiced in past centuries".

[TheHostPros at www.rense.com, 11-21-2001 and *Sovetskaya Rossiya*]

CAROL A. BROWN, professor, University of Massachusetts, in 1992, wrote to *Time* magazine:

"Last week I taught my students about the separation of church and state. This week I learned that the Pope is running U.S. foreign policy. No wonder our young people are cynical about American ideals."

This was in response to the February 24, 1992, article:

"Holy Alliance: How Reagan and the Pope Conspired to Assist Poland's Solidarity Movement and Hasten the Demise of Communism."

(Recall, in 1302, Boniface VIII allegedly stated: "It is absolutely necessary for the salvation of every human creature to be subject to the Roman Pontiff.")

[Saussy,F. Tupper. *Rulers of Evil.* New York. HarperCollins, 1999]

ANTONY C. SUTTON, in his 1995 book, *The Federal Reserve Conspiracy*, stated:

> *"Congressional passage of the Federal Reserve Act in December, 1913, must count as one of the more disgraceful unconstitutional perversions of political power in American history."*

DAVID MCCULLOUGH, renowned historian, author and speaker, Yale graduate, winner of three honorary degrees and the Pulitzer Prize, is quoted as saying in 1995:

> *"I'm convinced that history encourages, as nothing else does, a sense of proportion about life, gives us a sense of how brief is our time on earth and thus how valuable that time is."*
>
> Compare George Santayana, quoted about in 1905.
>
> [Reader's Digest article, "Why History?", December 2002]

WILLIAM JEFFERSON CLINTON, U.S. President, stated on March 31,1999:

> *"The worst thing you can do in life is underestimate your adversary."*

PAT SHANNON, journalist, in his forward to the 1999 book, *"Rulers of Evil,"* by F. Tupper Saussy, stated:

"The only people in the world, it seems, who believe in the conspiracy theory of history and those who have studied it."

///

GEORGE SOROS, "international financier", stated:

"Perhaps the greatest threat to freedom and democracy in the world today comes from the formation of unholy alliances between government and business. This is not a new phenomenon. It used to be called fascism...The outward appearances of the democratic process are observed, but the powers of the state are diverted to the benefit of private interests."

[Soros,George. *Open Society: Reforming Global Capitalism.* New York. Public Affairs, 2000. Page xi]

///

BILL MOYERS:

"Secrecy is the freedom zealots dream of; no watchman to check the door, no accountant to check the books, no judge to check the law. The secret government has no constitution. The rules it follows are the rules it makes up."

[Marrs,Jim, 2000. Moyers quoted on page 19]

///

ERIC J. CHAISSON, in his book (Cosmic Evolution):

"Looking forward toward the future, we search for enhanced understanding, for meaning and rationality. Is humankind part of a cosmological imperative, heading, perhaps, with other sentient beings, toward some astronomical destiny? Put bluntly...the scenario of cosmic evolution grants us unparalleled 'big thinking,' from which may well emerge the

global ethics and planetary citizenship likely needed if our species is to remain part of that same cosmic evolutionary scenario."

[*Cosmic Evolution: The Rise of Complexity in Nature.* Cambridge, Massachusetts, and London. Harvard University Press, 2001. See page 24]

%

DAVID C. KORTEN:

"The era of colonizing open frontiers is now in its final stage. The most readily available frontiers have been exploited, and the competition for the few that remain in such remote locations as Iran, Java, Indochina, Papua New Guinea, Siberia and the Brazilian Amazon is intensifying. There is now a vast literature and much debate on assessing the data…whether a particular limit has been exceeded or will be passed with the next few years. Such exactness is far less important than coming to terms with the basic truth that we have no real options other than to recreate our economic institutions in line with the reality of a full world."

[Korten, David. *When Corporations Rule the World - Second Edition.* San Francisco. Berrett-Koehler Publishers, 2001. Pages 35-36]

%

THOMAS L. FRIEDMAN, author and journalist, on CBS News, Sunday, June 30, 2002:

"There is nothing like telling the truth."

In April 2001,in a debate with Greg Palast in Cleveland, OH, he is quoted:

"It is increasingly difficult these days to find any real difference between ruling parties in those countries that have put

on the Golden Straitjacket... be they led by Democrats or Republicans, Conservatives or Laborites, Christian Democrats or Social Democrats."

(Golden Straitjacket meaning "global capitalism.")

[Palast,Greg; Shykles, Oliver. *Burn the Olive Tree, Sell the Lexus.* Article in *Everything You Know is Wrong.*Russ Kick (Editor). New York. The Disinformation Company, Ltd., 2002]

%

SMEDLEY D. BUTLER, Marine Major General (1881-1940):

"War is a racket ... War is largely a matter of money. Bankers lend money and send Marines to get it."

(Year unknown.)

[Marrs,Jim. *Rule of Secrecy.* See *Part II: The Finger- prints of Conspiracy.* New York. HarperCollins, 2000]

%

RICHARD METZGER:

"While the credibility of government sponsored "news" and information has long been considered suspect by both right and left, we've recently sensed a burgeoning perception by middle-of-the-road, mainstream Americans that 'disinformation' is everywhere — medicine, science, finance, commerce, media — often sponsored by very complex webs of power and influence."

[Also see Preface to *Everything You Know is Wrong.* New York. The Disinformation Company, Ltd - 2002. www.freespeech.org]

REVIEWS

One Moment in Eternity – Human Evolution is a sharp-tongued warning that religion is "holding hostage" science and scientific understanding of the world, particularly with regard to evolution. Medical doctor and former member of the American Academy of Psychiatry and the Law, Eugene Minard presents a stinging indictment of humanities atrocities in the name of religion, and a caution that our future as a species may very well result in extinction due to overpopulation, and its many consequences , as well as natural disasters.

The highest hope humanity has for its future includes separation of church and state. A passionate and persuasive discourse.

[Reviewed by Midwest Book Review, Oregon, Wisconsin]

Are humans still evolving ? Not noticeably in the physical sense, but culturally we certainly are. To that effect, are we becoming better, or worse ? After reading Doctor Eugene Watkins Minard's book, "One Moment in Eternity – Human Evolution", readers will certainly have something to think about.

An abundance of information awaits within these pages. Such research and thought that went into this work is astounding.

Beginning at the very beginning, with the origin of the known universe, and finishing with thoughts on the survival of civilization as we know it, Minard leaves no stone unturned in his quest for the truth. From what religious organizations have done to influence the beliefs of people of the world to what motivates our highest political leaders. From the very beginnings of life to the evolutionary process of Homo Sapiens. From Holy Wars to secret societies, Minard has thoughts and documented evidence to back up his ideas.

[Reviewed by Heather Froeschl]